GREEN ROOFS

Ecological Design and Construction

Earth Pledge

Schiffer Publishing Ltd®

4880 Lower Valley Road, Atglen, PA 19310 USA

Green Roofs: Ecological Design and Construction
is a project of the Earth Pledge Green Roofs Initiative

Publisher
Schiffer Books

Earth Pledge Executive Director
Leslie Hoffman

Earth Pledge Green Roofs Initiative
Colin Cheney

Managing Editor
Marisa Arpels

Editor
Siena Chrisman

Designer
Heather Sommerfield

Contributing Editor
Joel Towers

Assistant Editors
Elana Berkowitz, Gaby Brainard, Laura Hickey

Library of Congress Cataloging-in-Publication Data:

Earth Pledge.
  Green roofs : ecological design & construction / by Earth Pledge.
    p. cm.
  ISBN 0-7643-2189-7 (hardcover)
1. Green roofs (Gardening) 2. Buildings—Environmental aspects. I.
Title.

SB419.5.E29 2005
635.9'671—dc22
                                        2004025679

Type set in Helvetica

ISBN: 0-7643-2189-7
Printed in China

Published by Schiffer Publishing Ltd.
4880 Lower Valley Road
Atglen, PA 19310
Phone: (610) 593-1777; Fax: (610) 593-2002
E-mail: Info@schifferbooks.com

For the largest selection of fine reference books on this and
related subjects, please visit our web site at
www.schifferbooks.com
We are always looking for people to write books on new and
related subjects. If you have an idea for a book please contact
us at the above address.

This book may be purchased from the publisher.
Include $3.95 for shipping.
Please try your bookstore first.
You may write for a free catalog.

In Europe, Schiffer books are distributed by
Bushwood Books
6 Marksbury Ave.
Kew Gardens
Surrey TW9 4JF England
Phone: 44 (0) 20 8392-8585; Fax: 44 (0) 20 8392-9876
E-mail: info@bushwoodbooks.co.uk
Free postage in the U.K., Europe; air mail at cost.

GREEN ROOFS

Ecological Design and Construction

Earth Pledge

GREEN ROOFS

Ecological Design and Construction

Earth Pledge

Contents

Municipal Case Studies
Imagining the City: Urban Ecological Infrastructure | by Joel Towers

Appendix 1

Appendix 2

Earth Pledge Kitchen Garden
Earth Pledge Executive Director Leslie Hoffman, an avid gardener, became interested in green roofs during the construction of Earth Pledge's Kitchen Garden. The green roof is part of a larger renovation of a 1902 Georgian townhouse, which now serves as the Earth Pledge offices and a showcase for sustainable design.

Preface

By Leslie Hoffman | Executive Director, Earth Pledge

In the spring of 2002, I planted one of the first green roofs in New York City on the Earth Pledge townhouse in midtown Manhattan. Having been a designer and builder focused on green building in the 1980s and a passionate gardener for all of my adult life, I was pretty sure that I understood the reasons to install this roof. It would support our initiatives in sustainable agriculture and architecture, in addition to demonstrating an innovative technology that facilitates the transition to sustainability.

With rake in hand, spreading the growing medium over the layers of drainage, water retention board, and root barrier, I realized that the potential of green roofs was even greater than I had imagined. Green roofs represent an elegant opportunity to simultaneously mitigate environmental problems and create immediate life-enhancing value. They individually offer building owners savings on energy and roof membrane replacement costs, while also greening the cityscape for owners and residents of neighboring buildings. Flowering and native plants help cool the urban landscape and combat the pollinator crisis in our region, and one doesn't need to hear much about combined sewage overflow and the erosion and runoff issues in coastal zones to understand why pervious surface is desirable. But perhaps the most overlooked incentive for urban green roof development, and one that will likely take on greater significance in coming years, is the connection to the earth—to soil, plants, water, and nature—that green roofs offer to city-dwellers—people who are often disconnected from the very source of their own survival. That this connection could be provided by such a simple layered system on the otherwise wasted resource of vacant roof space is what I had not fully imagined.

We created the Green Roofs Initiative at Earth Pledge to research, educate, and demonstrate how cities can meet the environmental challenges we face—while embracing this new infrastructural element as a tool to inspire urban dwellers to reclaim their long-forgotten connection to nature.

Green roofs are one piece in the puzzle of creating the ecological cities of the twenty-first century. According to the United Nations, 2005 is the first year in history that more than half of the world's population will live in cities. This is a trend that we must adapt to, just as we begin to grapple with the impacts of our industrial practices and global climate issues. At the center of all Earth Pledge initiatives is a focus on re-imagining how our urban centers function in relation to natural systems.

Founded in 1991, Earth Pledge works to explore replicable solutions that inspire and facilitate a global transition to sustainability. We work with farmers, chefs, and institutional food purchasers through our Farm to Table Initiative, developing new ways to reduce pesticides and petroleum-based fertilizers in agriculture and creating new markets for local, sustainable farm products. The anaerobic digestion technology promoted by our Waste=Fuel Initiative is an important alternative to conventional waste management, which works by breaking down organic materials in the waste stream into usable biogas. Our Brownfields Initiative has created The Guardian Trust as a way to ensure that redeveloped brownfields are monitored to protect the environment and human health long into the future.

The Earth Pledge Green Roofs Initiative works with the many sectors necessary to achieve widespread green roof infrastructure. Since 2001, we have educated hundreds of building and design professionals. We convened a task force of fifteen government agencies to evaluate options for public policy support of green roofs. It became clear that an understanding of the impacts of green roofs would enable policy makers to make sound decisions. In response, we created a multidisciplinary research project, the New York Ecological Infrastructure Study, to quantify the value of green roofs. GreeningGotham.org, our resource for green roof information, has garnered support from Mayor Michael Bloomberg, Senator Hillary Clinton, the Speaker and numerous members of the City Council, the Regional Administrator of the Environmental Protection Agency, City Commissioners, actor Ed Norton, Robert Kennedy, Jr., and many others.

Green Roofs: Ecological Design and Construction draws together in one volume the principles of ecological green roof design and construction that Earth Pledge has fostered. From the words of William McDonough, whose visionary essay opens the book, to the exemplary ecological green roof projects in cities around the world, this book presents an understanding of green roofs as a tool to transform our urban landscape into a natural system that can support and sustain us. I hope this book inspires you to join the growing worldwide movement to build and implement widespread green roof infrastructure.

Introduction

On this rooftop...
I'm watching you
move among your sparse,
pinchpenny flowers,
...that pull the sun's rays in
as best they can
and suck life up from one
mere inch of dirt.

Howard Moss, "The Roof Garden"[1]

As in Howard Moss's poem, rooftop gardens of New York City are a landscape where residents challenge the stark, "needle hardness" of the city. Gardens of varied sizes and designs have long dotted the city's roofscape, reframing New Yorkers' experience of their city by drawing nature into the urban landscape.

In the twenty-first century, these rooftops are being re-imagined once again, this time as the theater for environmental problem-solving. In cities across the world, "green roofs" are being implemented as an effective and attractive response to such urban challenges as stormwater runoff pollution and high temperatures. While modern green roofs require the creative and technical skill used to design traditional roof gardens, their success necessitates that they be considered as part of the wider human and natural community in which they are built.

Green Roofs: Ecological Design and Construction was created as a resource to showcase, inspire, and guide the creation of environmentally-conceived roof gardens. This is the first book on roof gardens to explicitly frame design and construction in an ecological context. The compilation of essays and case studies aims to

COURTESY OF EARTH PLEDGE

educate the reader about the benefits and design possibilities of modern green roofs at both the building and municipal levels. In addition, this book will address how a single green roof is part of a larger ecological network with environmental and health benefits, and how green roofs can be designed to capitalize on this important role.

The book is organized to allow the reader to move from the vision and rationale for environmentally-conceived green roofs, to case studies of exemplary projects, and finally to a look at the various approaches cities have used to develop green roof networks.

The first two chapters establish green roofs in the context of sustainable design and environmental problem-solving. Visionary green architect William McDonough writes in the first chapter, "the integrative design [of a green roof] … enables the building and its inhabitants to *participate* in natural processes, allowing for an appreciation of the relationship between creativity and the abundance of nature." The second chapter, "From Grey to Green: Environmental Benefits of Green Roofs," presents the details of how green roofs are used to address such

environmental challenges as the urban heat island effect, stormwater runoff pollution in urban waterways, resource use, disappearance of native species, and human health.

The subsequent section presents 40 case studies of green roof projects that demonstrate the wide variety of ways in which green roofs can respond to environmental and social issues while demonstrating exceptional innovative design. In his essay preceding the final section of the book, architect Joel Towers outlines how green roof design can be translated from the building level into what he frames as "urban ecological infrastructure." The final section presents seven case studies of municipalities exploring large-scale green roof development. The experiences of Berlin, Tokyo, London, Portland, Chicago, Toronto, and New York offer distinct narratives on pursuit of green roof infrastructure, key players, government participation, and success. The appendices offer an overview of green roof system components and a detailed outline of the design details for all of the building case studies in the book.

The ecological design and construction of green roofs offers a new way for city residents around

the world to bring nature into their daily lives. With environmental challenges becoming more apparent everyday, new solutions are required to address these problems elegantly and effectively. It is our hope that this book can lead you to consider how building owners, residents, and municipalities can solve some of these challenges while experiencing the life-enhancing value of being surrounded by gardens. In learning the principles of an environmentally functional green roof, you will become at once more attuned to the problems our cities face and to how growing flowers from a "mere inch of dirt" atop your roof can help improve the ecology of the city.

Earth Pledge Kitchen Garden
Located in midtown Manhattan, the Earth Pledge demonstration green roof is planted with over sixty species of sedums, flowers, and edibles, including corkscrew rush, alpine strawberries, and gold nugget tomatoes.

A Field of Dreams
Green Roofs, Ecological Design and the Future of Urbanism
By William McDonough

I am strolling in a field listening to crickets and watching birds pluck insects from the dirt. Wildflowers bend in the wind. Warblers and thrushes flit about in tall native grasses and soar over the rolling terrain. The scene is rich, beautiful, and lively—some might say wild. But this landscape is also a cultural space; I am standing on top of a building.

The building, the centerpiece of Gap Inc.'s corporate campus in San Bruno, California, is a pioneering office building with a green roof that is more than just a pretty patch of sod. Blanketed in soil, flowers, and grasses, the roof's hills echo the local landscape, reestablishing several acres of the surrounding coastal savannah ecosystem. The native plants and soil also absorb stormwater, filter the air, and provide thermal and acoustic insulation. From inside the building, one can look out the window at the rooftop grasses or enjoy a fresh summer breeze. In these and many other ways, the roof incorporates the landscape into the design of the building.

In addition to the obvious practical considerations such as cost and scheduling, we began the Gap project by asking a range of design questions not often considered by architects, planners, and their clients. "What would make employees look forward to coming to work in the morning?" and "What would native birds hope to see as they fly over the site?" Questions like these change the nature of the design process, expanding its concerns into disciplines such as ecology, botany, conservation biology, and environmental history, offering a lens through which one can see the natural systems at work—landforms, hydrology, vegetation and climate of a particular locale. Thoughtfully applied, they empower architects and planners to develop designs that "fit"; designs that encourage healthy and creatively interactive relationships between a building and its environs. In other words, the human impact on the environment can be positive, vital, and good—even regenerative. As this idea takes root in the world of planning and architecture, it offers hope for a mutually-enriching relationship between nature and human culture, as well as a fresh, inspiring direction for urban design.

Technology and performance play critical roles in the pursuit of ecological intelligence. The design and construction of green roofs and buildings demands an extraordinary range of technical expertise, from understanding stormwater hydrology or the flux of ultraviolet radiation to constructing an effective rooftop waterproofing system. What is crucial is a design approach that accounts for a wide variety of economic, environmental, and cultural criteria. When such an approach is supported by technological know-how, a truly delightful, high-performance building can result.

Consider again the Gap building. The green roof is one of several integrated systems designed to create a productive and comfortable workplace. While the rooftop soils and grasses insulate the building from the midday sun and the sound of jets flying overhead, a raised-floor cooling system allows evening breezes to flush the building at night. The windows can be opened for fresh air and daylight provides natural illumination. There are indoor and outdoor public gathering spaces, which are enlivened by artwork, plants, and a cafe. It is a wonderful place to go to work. And passing birds do not see a flat black tarmac, but a rolling flowering grassland.

All that, and the Gap building is also one of the most energy efficient buildings in California. By setting out to create a positive, regenerative human footprint, by tapping local energy flows and integrating construction and landscape, the building outperforms those that set energy effi-

901 Cherry Street, Offices for Gap Inc
The Gap office building in the San Francisco hills blends almost seamlessly into a steeply sloping site, with its undulating roofline echoing the surrounding terrain. The 69,000 square foot green roof, covered in native grasses and wildflowers, reproduces the local coastal savannah ecosystem.

COURTESY OF MARK LUTHRINGER/ WILLIAM MCDONOUGH + PARTNERS

ciency as their highest goal. The integrative design also enables the building and its inhabitants to *participate* in natural processes, allowing an appreciation of the relationship between human creativity and the abundance of nature.

Imagine that sensibility alive in our cities. Imagine cities from New York to Los Angeles tuning in to the natural processes at work in the urban world. For most of the last 150 years, urban nature has been synonymous with urban parks—not the city itself. Even as Frederick Law Olmsted's landscapes naturally reduce urban flooding and improve air and water quality, just as he imagined they would, few urban dwellers see his parks as anything more than ornamental respites from the harder urban world. They are loved, but separate.

There is another view. In the words of landscape architect Anne Spirn, "The city is a granite garden, composed of many smaller gardens, set in a garden world...The city is part of nature." Nature is in the air we breathe, the earth we stand on, the water we drink, and the organisms with which we share our habitat. Nature in the city is a powerful force that can shake the earth and crumble buildings. It is the millions of fossilized

organisms cemented in the limestone of a downtown building. It is rain and the rushing of underground rivers buried in storm sewers. It is the water from the faucet, delivered by pipes from an outlying river or reservoir, used, and then washed back to the waters of the river or the sea. Nature in the city is an evening breeze whispering down a side street, an eddy swirling in the gutter, the patterns of sun and shadow on the sidewalk. It is the natural processes that govern the transfer of energy, the movement of air, the erosion of the earth, and the hydrologic cycle. The city is part of nature.

With this idea of the natural city in mind, architects and planners have begun to integrate natural processes into urban life. We see the signs in unearthed urban rivers, reforested riparian corridors, and reclaimed wetlands purifying the water supply. New building and street designs take advantage of natural cooling air flows. Energy is being collected from rooftop solar panels and from underground geothermal sources. In community gardens, citizens have daily interactions with soil, water, and living things. Green roofs are filtering stormwater, easing the heat island effect, and providing habitat for native plants, birds, and insects. A partnership is strengthening between

nature and the city that has the potential to create a life-affirming urban realm.

Green roofs are a key element of this transformation. Imagine, for example, the effect of community kitchen gardens on the rooftops of New York City. Neighborhoods might become urban agricultural districts; the growing of food providing sustenance, a relationship with nature and neighbors, and an opportunity to enjoy the fruits of one's labor. The gardens would make visible the vital connections between water, soil, food, and human culture, and create a network of living landscapes across the ancient archipelago that is New York City. Add to that the energy savings a sea of green roofs might provide with their heating and cooling benefits—according to *The New York Times*, those savings could amount to as much as $16 million a year.

How better to signal a sea change in the priorities of local government than a green roof on City Hall? That is exactly what happened in Chicago. Mayor Richard Daley has proclaimed that Chicago will become "the greenest city in America." As his administration plants hundreds of thousands of trees, restores the Lake Michigan shoreline, and invests in renewable energy, the green roof on Chicago's City Hall has become a symbol of the city's commitment to change.

It is a change worth watching. Mayor Daley not only *says* he wants the city to be the greenest in America, he is making environmental initiatives an integral part of a long-term strategy for growing economic and social health. To that end, says Department of the Environment First Deputy Commissioner David Reynolds, the city is working "to bring industry back to Chicago while also revitalizing local ecology." The mayor is committed, he says, to "making the city a national model of how industry and ecology can exist side-by-side."

The work is already underway. Along with widespread greening initiatives, the city has undertaken the largest brownfield redevelopment effort in the United States. It has committed to buying twenty percent of its electricity—for schools, libraries, subways, and streetlights—from renewable sources by 2006. Meanwhile, renewable energy companies, such as the solar panel manufacturer Spire, have moved their headquarters to the Chicago Center for Green Technology, a new ecologically intelligent facility built on a restored industrial site. Spire is already supplying Chicago with locally manufactured solar panels, which the city has installed on the roofs of the Field Museum, the Mexican Fine Arts Museum, and the Art Institute of Chicago.

Unfortunately, these positive changes in our cities could end up being a flash in the pan unless they are placed within the context of a sustaining, long-term vision. A green roof is a wonderful addition to a city neighborhood, but its impact grows when it is conceived as a humble first step toward a deep revitalization of urban life. In other words, urban design must be strategic rather than piecemeal,

with each initiative supporting the goals of a holistic, integrated plan. This is not news in Chicago. The city government is developing a set of urban planning principles to guide decision-making over the long haul, not just during the Daley Administration, but well into the future.

David Reynolds put it this way: "We have been saying that we are going to be the greenest city in America. But to truly become a thriving green city we need to carefully define what that means and what we should be striving for, day-by-day and year-by-year. No city in the United States has really gotten this right yet, and we believe that part of the problem has been that no American city has developed a set of guiding green principles—akin to the timeless principles of the Constitution—that describes its ideals, sets its course, and defines its means. That's what we are trying to do in Chicago. And we hope the principles we develop become so well known and so well understood that they define how we operate as a city government for the next one hundred years."

The fruits of this labor, the soon-to-be-published Chicago Principles, developed with William McDonough + Partners, will serve as a reference point for the city as it pursues Mayor Daley's dream. The Principles, it is hoped, will provide a coherent, ecologically intelligent foundation for urban design. They support strategic decision-making and encourage planning choices that enhance not only environmental health, but economic productivity and social welfare as well. The Principles share the spirit and aim expressed in the nine declarations of the Hannover Principles, which my colleague Michael Braungart and I crafted for the city of Hannover, Germany, in 1992. The Hannover

Principles establish a number of tenets including: insisting on the right of humanity and nature to co-exist in a healthy, supportive, diverse, and sustainable condition; the essential interdependence between elements of human design and the natural world; committing to the elimination of the concept of waste; increased reliance on natural energy flows. The Chicago Principles similarly hold the promise of a road towards an urban future that is restorative and regenerative by design.

In our work over the past decade, my colleagues and I have found that following these guiding principles in everything we do—from designing buildings and community plans to products and factories—yields extraordinary results. The green-roofed Gap building is one example of this approach. Another is the revitalization of Ford Motor Company's 1.2 million square foot Rouge River manufacturing complex in Dearborn, Michigan, where we helped envision the largest green roof in the world.

The Rouge River restoration illustrates the benefits of considering a diverse range of economic, social, and ecological concerns in urban design. As we approached the design process with Ford—which had decided to invest $2 billion over 20 years to transform the Rouge into an icon of twenty-first century industry—people wondered if a blue chip company with a sharp focus on the bottom line could take a step toward something truly new and inspiring. Could environmental restoration and profits co-exist?

The answer is *yes*. In fact, for this agenda to become widespread, they must. Using the Hannover Principles as a guide, and working in

Ford River Rouge, Truck Plant
The green roof is projected to retain fifty percent of the annual rainfall—447,000 gallons of water annually—for $10 million less than conventional water treatment systems. It is also expected to decrease energy consumption by seven percent and improve air quality by forty percent by absorbing dust and breaking down hydrocarbons.

Ford River Rouge, Truck Plant
Bigger than eight football fields, the Ford assembly plant green roof is the largest in the world. The company is further greening the factory site with lawn space, trees, trellises, and a wetland.

COURTESY OF FORD PHOTOGRAPHIC/ WILLIAM MCDONOUGH + PARTNERS

collaboration with Ford's executives, engineers, and designers, we explored a variety of innovative ways of creating value. Rather than using conventional economic metrics to reconcile apparent conflicts between environmental concerns and the bottom line, Ford's leaders examined how smart design decisions could grow social and ecological value as well as profits.

The results were inspiring. Instead of trying to meet its environmental responsibilities as cost-effectively as possible, Ford opted for a manufacturing facility that would create habitat, make oxygen, connect employees to their surroundings, and invite the return of native species. The new plant features skylights for lighting the factory floor and a 10-acre roof of soil and growing plants. The living roof provides habitat for birds, insects, and microorganisms and, along with porous paving and a series of constructed wetlands and swales, will absorb and filter storm water runoff for $10 million less than conventional water treatment systems. In addition, native grasses and other plants are ridding the soil of contaminants and a variety of trees are being planted to aid in the bioremediation. This is a landscape of renewal.

The Rouge restoration has important implications for cities and urban designers. It shows not only that green roofs and other ecological technologies can be effectively and profitably used in large-scale urban projects, but also that industry and ecology can flourish side-by-side.

With this in mind, we can begin to imagine cities engaging ever more creatively with nature.

This is possible when we design each thing we make as a nutrient that can circulate in safe, regenerative closed-loop cycles—a biological cycle in which organic materials are returned to the soil; a technical cycle in which high-tech, synthetic materials are produced, used, recovered, and remanufactured. An upholstery fabric I designed with Michael Braungart, for example, can be tossed on the ground to nourish the soil when it wears out; in the city it could become food for rooftop gardens. Similarly, high-tech products, such as perpetually recyclable fibers or windmill blades, can be "food" for technical systems, providing safe materials for generations of useful goods. These *cradle-to-cradle* material flows (as opposed to the typical cradle-to-grave flow of materials from producer to consumer to landfill) are crucial to sustainable urban design. This kind of cycle ensures that the materials with which we build our cities will be healthful and beneficial, eliminating the concept of waste, and providing a clean, productive economic base for healthy urban growth.

And so we can begin to see the city not only as an elegant self-sustaining place, but as a revitalizing force in its region. In this new metropolis, biological and technical nutrition flow between city and countryside and enrich both. The city receives food, water, and energy from a broad nexus of solar-powered, biologically-based, photosynthetic

systems. The energy of the sun is harvested on rooftops; rural windmills power city buildings; water flows from green roofs into the soil and the watershed. In the countryside, farmers grow food using implements manufactured in the city—technical nutrients—and the city receives this nourishment, digests it, and then excretes it back to its source, returning biological nutrients to the rural soil. The farm's windmills are forged in the city; produce power for the region in the countryside; are returned to the city for periodic refurbishment; and then returned to the farm. Everything moves in regenerative cycles: from city to country and country to city; all the polymers, metals, and synthetic fibers flow in the technical cycle; all the photosynthetic nutrients—food, wood, natural fibers—flow in the biological cycle. These cycles of nutrients are the twin metabolisms of the living city that allow human settlements and the natural world to thrive together. If we are to make our cities truly sustaining and sustainable, we need to make this vision a strategic truth that informs all of our designs.

Our vision, simply put, is this: A world of interdependent natural and human systems, powered by renewable energy, in which everything we make flows in safe, healthful biological and technical cycles, elegantly and equitably deployed for the benefit of all. The view from the rooftop suggests that this dream is within our grasp and, indeed, that it has already taken root in the granite gardens of our garden world.

From Grey to Green

Environmental Benefits of Green Roofs

By Earth Pledge Staff and Katrin Scholz-Barth

COURTESY OF OPTIGRÜN AG

The dense, accelerated pace of modern urban development has affected many of the earth's natural processes. Asphalt and concrete rooftops, roads, and parking lots cover up to seventy percent of land area in dense cities like New York, while open space in sprawling cities like Phoenix, Arizona is lost to development at a rate of 1.2 acres per hour.[1] Green spaces—open areas with soil and growing vegetation—mediate temperatures, absorb stormwater, slow runoff, and support biodiversity and human health. In order to restore balance to urban ecosystems, cities must find ways to restore green space to an increasingly grey world. But green space offers less economic value than other land use, like parking lots or buildings. Environmentalists in inner cities, where land is scarce and economic pressure intense, must think creatively, plant-

ing trees and gardens on traffic islands and along streets. The many acres of flat rooftop space in most cities present another possibility. A city's bare, black rooftops can become new green space—without altering land use or compromising development.

Green roofs are lightweight, layered systems that cover conventional roof surfaces with growing medium and plants. The simplest green roofs (*extensive*) are shallow; 3 or 4 inches of growing medium planted with drought-tolerant succulents or grasses and requiring minimal maintenance. Deeper, more elaborate green roofs (*intensive*) can be landscaped with flower and vegetable gardens, or even trees. Roof plantings have many historic precedents, from the hanging gardens of Babylon to Scandinavian sod roofs. Today's

green roofs are based on German designs from the 1970s and use a lightweight, mineral-based growing medium to support plant growth. A citywide or regional network of green roofs has the potential to address some of the most serious environmental problems facing cities in this century. Green roofs challenge traditional perceptions of cities as places apart from nature by integrating living plants into buildings.

Waldspirale

The architect Friedensreich Hundertwasser (1928-2000) was a critic of the modern architect's emphasis on rationalism, artifice, and human superiority. His own designs, such as the Waldspirale in Germany, reflect a vision of an architecture based on natural forms and emphasizing vegetation.

Climate

Cooling Cities with Green Roofs

New York City
Temperatures are higher in high-density cities like New York than in surrounding areas due to the concentration of heat-radiating surfaces and the lack of vegetation—a phenomenon known as the urban heat island effect. Green roofs cool and filter air, help reduce the urban heat island, and can mitigate overall climate change.

125 Maiden Lane
The rooftop of 125 Maiden Lane, in New York City's financial district, combines low albedo surfaces and greenery to lower summer cooling costs and create a relaxing respite for hot days.

COURTESY OF EARTH PLEDGE

COURTESY OF CHROMA

During the summer, urban areas are often 2°F to 8°F hotter than surrounding areas, a phenomenon known as urban heat island effect.[2] The intensity of this gradient varies, depending on climate, topography, and urban design. New York City is typically 3.6°F to 5.4°F warmer than its suburbs.[3] The effect is more pronounced in tropical cities, often reaching 16°F in Mexico City.[4] As global warming and urbanization trends continue, heat islands will grow. Already, urban areas are heating up faster than the earth as a whole: Japan's average temperature has increased by about 1°F over the past century, but the temperature in Tokyo has increased by 5.2°F—over five times as much.[5]

Heat islands are created when dark-colored, impermeable surfaces absorb heat energy and radiate it back into the air. The amount of energy that a surface reflects, which determines how hot it will become, is called albedo, measured on a scale of from 0 to 1 (hottest to coolest).[6] The albedo of a tar or gravel roof is about 0.08, as compared to 0.25 for grass and 0.6 for reflective roofing.[7] Asphalt and concrete absorb and re-radiate the most heat, thus the prevalence of asphalt and concrete rooftops plays a major role in the heat island effect.[8] Green roofs, on the other hand, are not only

more reflective than black roofs, but include plants that actively cool the air. Plants stay cool by drawing moisture from the soil and evaporating it through their leaves, a process called evapotranspiration. This process cools the leaf surface and the surrounding air.

Urban heat islands have serious consequences for the environment and human health. Higher summer temperatures increase electricity demand. In large U.S. cities, peak summer utility loads are estimated to rise one and a half to two percent for each 1°F increase in temperature.[9] Even slight temperature increases could cost ratepayers billions annually and require costly new generating facilities.[10] Air quality declines as temperatures rise; power plants emit more pollutants to keep up with energy demand, while smog and dangerous pollutants form more readily at high temperatures. The number of days when air pollution levels exceed federal standards could increase by ten percent for each 5°F temperature increase.[10.5] Heat-related illnesses like heatstroke and respiratory illness also increase, particularly among the most vulnerable inner city populations: low-income residents, the elderly, and young children.[11] Chicago's 1995 heat wave caused 739 deaths in five days.[12]

There are two ways to mitigate urban heat island: increasing vegetation, or increasing surface reflectivity. Green roofs accomplish both, and considerably reduce individual building energy use. While a typical asphalt roof can reach 160°F on a summer day, green roofs and other vegetated surfaces rarely exceed 80°F.[13] Evaporative cooling reduces heat transfer through the roof into the building, making the inside cooler and reducing the need for air conditioning. The National Research Council of Canada found that green roofs reduced the average daily energy demand in a 400 square foot test facility by over seventy-five percent.[14] An Environment Canada study suggested that greening a minimum of six percent of total available roof space (70 million square feet, or 6.5 million square meters) in Toronto could reduce summer air temperature by 1.8°F to 3.6°F (1°C to 2°C).[15] The study also showed that the subsequent decrease in energy use would further lower temperatures by an additional degree. It estimated that a 1.8°F (1°C) temperature reduction would result in a five percent reduction in cooling demands of all buildings. The synergetic impact could considerably reduce overall urban heat island effect.

Hydrology

Managing Stormwater with Green Roofs

In addition to raising temperatures, urban development disrupts the natural movement of water, known as the hydrologic cycle. Precipitation cannot infiltrate asphalt and concrete, and instead it runs off, potentially overwhelming city sewage systems. The volume of this runoff is immense: greater Atlanta processes between 57 and 133 billion gallons of stormwater runoff annually—equivalent to the annual household water needs of its entire population.[16]

The impact of stormwater runoff on water quality depends on the sewer system that manages it. Many older cities have combined sewage systems, which channel household sewage, commercial and industrial wastewater, and stormwater runoff into a single wastewater treatment plant. During rainstorms, runoff dramatically increases the volume of water in the system. This flood of wastewater can overwhelm the treatment plant, causing overflow to spill untreated into waterways—a phenomenon called combined sewage overflow (CSO).[17] CSO discharges contain pathogens, toxins, nutrients, and many other pollutants that endanger human health, triggering swimming restrictions and fishing bans.[18] Modern sewer systems avoid CSO by separating sewage and stormwater. Sewage is sent to treatment plants, while stormwater is discharged directly into receiving water bodies. This eliminates CSO, but stormwater runoff still carries toxic contaminants from streets and sidewalks. Further, nutrients in runoff can cause explosive plant growth in receiving water bodies, including toxic algae blooms ("red tides") that choke marine life.

Almost 800 communities in the United States—serving 40 million people—have combined sewer systems. Combined sewer overflows happen with alarming frequency.[19] In New York City alone, 40 billion gallons of untreated wastewater—twenty percent of which is raw sewage—are spilled into local waterways each year from the CSO system.[20] In Portland, Oregon, local streams are highly polluted, and salmon populations are endangered. Beach closures in Chicago made headlines in 2001, as sewer overflows caused E. coli contamination in Lake Michigan.

The Clean Water Act mandates that cities curb CSO and stormwater runoff pollution. Many proposed solutions—sewer separation, underground storage tanks, upgrades to wastewater treatment facilities—require massive public investment in new infrastructure. The estimated cost of structural modifications to reduce CSO in Portland, Oregon, was estimated at $1 billion.[21] New York City has allocated $1.8 billion for CSO abatement projects, most of which will fund sewer upgrades and an underground retention tank.[22]

Green roofs offer an alternative. Vegetated rooftops retain and detain stormwater, reducing runoff volume and slowing the rate at which it enters the sewage system. Research by the Penn State Green Roof Research Center, Portland's Bureau of Environmental Services, and North Carolina State University found that extensive green roofs can retain fifty percent of a 1-inch rainfall and as much as fifty to seventy percent of rainfall annually. These studies also

found that green roofs can delay runoff from thirty minutes to four and a half hours, and slow it from forty-two to ninety-six percent.[23] Delaying runoff is as important as reducing its volume; it is the initial flood of stormwater that triggers overflows. Green roofs also act as filters, reducing the pollutant load delivered to waterways. Plants and soil trap airborne pollutants, and heavy metals bind to soil particles.[24]

Green roofs are a cost-effective stormwater management tool compared to conventional treatment and retention methods. The Environment Canada study of Toronto (cited above) found that by greening six percent of available roof spaces at a cost of $45.5 million (CDN), the city could retain as much stormwater (127.1 cubic feet per year or 3.6 million cubic meters per year) as a storage tank costing $60 million (CDN)—a savings of $14.5 million.[25] Many cities have begun to promote green roofs for stormwater management. Portland, Oregon, for example, has officially endorsed the technology by adding "ecoroofs" to its list of approved stormwater management approaches.[26]

Left: Penn State Test Plots
The Penn State Center for Green Roof Research has three buildings with green roofs and three with conventional roofs. The center is testing the effectiveness of green roofs at lowering heat flux, retaining stormwater runoff, and improving water quality.

Right: Bondorf
The German food distributor Bondorf installed a 12-acre green roof on its production and transportation facility in response to a federal mandate to manage rainwater onsite. The roof stores and evaporates at least seventy percent of annual rainfall, reducing site runoff by 6.25 million gallons per year.

COURTESY OF DAVID BEATTIE

COURTESY OF OPTIGRÜN AG

Urban Ecology
Preserving Wildlife and Building Habitat

Green roofs can support biodiversity and raise awareness of the link between the city and the natural world. Preserving biodiversity—a common measure of ecosystem health—is particularly critical in developed areas. Habitat fragmentation, pollution, and noise make modern cities hostile to most plant and animal species. Green roofs can create healthy, functioning habitat in the midst of the urban landscape.

Many green roof projects include native plant species. Chicago City Hall *(Building Case Study 18)* tests the survivability of over 150 native and non-native plants in a variety of soil depths and drainage conditions. Other green roofs replicate local ecosystems. The Sechelt Justice Services Centre in Sechelt, British Columbia, *(Building Case Study 07),* uses native grasses and sedums to mimic Puget Sound's coastal meadows. The roof of the Gap headquarters in San Bruno,

California, was also designed with native grasses.[27] Other projects go further, preserving endangered plant species on green roofs. In Mexico City, the nonprofit organization CICEANA cultivates 25 species from the threatened Pedegral de San Angel ecosystem on its green roof *(Building Case Study 39).* On a green roof in Nashville, Tennessee, landscape architect Eric Shriner planted the endangered Tennessee Purple Coneflower and 14 other species from the threatened Cedar Glade ecosystem.[28]

These projects are models for ecosystem restoration, but they are still constructed landscapes, and the plants selected are native to the region but rarely native to the site itself. An alternative approach preserves ecosystems that are already thriving on the specific building site. "Brown roofs" or "rubble roofs" use soil from the site itself, allowing it to be colonized by windblown

seeds and local fauna. The potential of brown roofs to preserve biodiversity is illustrated by the Orchid Meadow on the Moos water filtration plant near Zurich, Switzerland *(Building Case Study 40).* Swiss biologist Stephan Brenneisen's experiments at Basel University Hospital and the Institute for Hospital Pharmaceuticals *(Building Case Study 37 and 38)* demonstrate that brown roofs attract a variety of animal species, including beetles, spiders, and migratory birds.[29] New research in England has also shown that sedum-based roofs can become thriving habitats. In a study of green roofs in London's new financial center, Canary Wharf, Genevieve Kadas discovered over 59 species of spiders—two varieties were classified as nationally rare, six were new to London, and one was the first of its kind ever recorded in Southern England *(Municipal Case Study: London).*

Bottom left: N.E.U. Development
The Triangle Building at the N.E.U. Development in Nashville, Tennessee, is built on a former brownfield. The roof is planted with 1,200 plants from 15 species belonging to the endangered Cedar Glade ecosystem.

Above left and right: Ducks Unlimited National Headquarters and Conservation Center
The Ducks Unlimited National Headquarters and Conservation Center in Winnipeg, Canada, is built adjacent to the Oak Hammond Marsh, an important site for wildlife, particularly migrating waterfowl. The green roof was designed to help the building blend into the landscape. The roof is planted with native species and is essential to minimizing the impact of the development on the marsh.

Quality of Life
Enhancing Urban Well-Being

Birds and invertebrates are not the only species that thrive in greener cities. Several studies have linked the calming effects of green plants to shorter patient recovery times. Hospitals, including St. Luke's Science Center in Japan (*Building Case Study 17*) and Vancouver's General Hospital, have built roof gardens for use by patients. Greenery in cities improves the quality of life for urban residents, decreases stress, and creates space for relaxation and recreation.

The popularity of green spaces is also reflected in real estate value. Developers in Tokyo have begun to install elaborate roof gardens, which significantly increase the value of the building. Green roofs have other economic benefits as well. While normal roof membranes last 10 to 15 years, a green roof can double or triple the life of the roof membrane by eliminating weather-related expansion and retraction and damaging exposure to sunlight. Additionally, the insulation provided by a green roof lowers energy costs. William McDonough + Partners estimated that the green roof on the GAP Corporation's Headquarters in San Bruno, California, would pay for itself in 11 years based on the cost of the initial investment, soil and plants, structural upgrade, annual energy and operation savings, as compared with a conventional EPDM roofing system.[31]

The benefits of these savings more easily outweigh green roof costs in places like Germany, where the price of the technology is significantly lower. In the United States, where cost is high and few incentives exist, the economic benefits alone do not generally justify the investment in a green roof. Governmental involvement, in the form of education, incentives, and regulations, would lower barriers to adoption and stimulate the market.

Top: Ecover Company
The Belgian Ecover Company installed a green roof on its factory in Oomstalle for the cooling benefits. The factory has become a symbol of the company's ecological image and is featured on its detergent bottles.

Bottom left: Oaklyn Branch Library
The exceptional insulation and energy savings provided by the meadow on top of the Oaklyn Branch Library make the roof pay for itself. The extra initial expense of the native prairie grass roof was more than offset by the fact that three sides of the building are simple cast concrete retaining walls. The unusual site and building design allowed the library to be constructed for less than the proposed budget, while providing a building that will be ultra-energy efficient and easy to maintain for many years.

Bottom right: Fairmont Waterfront Hotel
According to the Fairmont Waterfront Hotel in Vancouver, British Columbia, its green roof garden produced $30,000 (CDN) of herbs and produce in a year.

COURTESY OF ARCHITECTEN ATELIER MARK DEPREEUW

COURTESY OF VEAZEY PARROTT DURKIN & SHOULDERS

COURTESY OF THE FAIRMONT HOTELS AND RESORTS

Ecological Infrastructure and Sustainable Cities
Building a Greener Metropolis

Federal policies like the Clear Air Act and Clean Water Act are designed to ameliorate major environmental problems. Much of the financial burden of the reforms, however, falls on municipalities. Some cities have begun to use greening strategies as a less expensive way to meet federal requirements. The jump from individual building to ecological infrastructure cannot be done without government legislation. Government demonstration projects, incentives, and regulation promote green roofs by offsetting costs and stimulating the market. The case studies of cities from Toronto to Tokyo demonstrate how municipalities around the world are encouraging green roof construction through legislation and funding.

In North America, the development of the Leadership in Energy and Environmental Design (LEED) green building standards has influenced government policy-making. LEED standards evaluate building projects on a 69-point system, which considers site development, water savings, energy efficiency, material selection, and indoor environmental quality. Buildings are certified silver, gold, or platinum according to the points of their environmental measures. Green roofs qualify for up to three points in the United States. The standards were developed to encourage the private sector to investigate the cost-saving and marketing potential of green building, but city governments are now using them as a benchmark for sustainable design. Many federal agencies, states, and cities require that new government-funded buildings meet LEED standards. Cities like Portland and Seattle also offer incentives for private-sector LEED buildings.

Green roofs represent a new trend in urban planning, integrating lost natural processes into man-made structures, technology working with nature instead of replacing it. Implementing a network of green roofs in metropolitan areas throughout the United States can be an important step towards making our cities sustainable, healthy places to live.

Left: Seattle Justice Center
The LEED platinum-rated Seattle Justice Center is one of many municipally-funded demonstration projects that aim to raise awareness about the benefits of green roofs and other green technologies.

Right: Aldi Warehouse
The complete coverage green roof on the Aldi warehouse in Murr, Germany, demonstrates the impact of German laws requiring green roof installation on greenfield developments.

COURTESY OF THE CITY OF SEATTLE

COURTESY OF XEROFLOR

Building Case Studies

A Framework for Green Roof Design and Construction

Green roofs offer a new paradigm for the conceptualization and construction of rooftop gardens. In their materials, form and function, the green roofs detailed in this book offer models for how rooftop gardens can meet private needs while being a positive, functional component of the local ecology.

Landscape architects have traditionally designed roof gardens solely as private amenities. By requiring heavy irrigation, herbicides and pesticides, or new plantings every season, many traditional roof gardens place a burden on their local environments. In contrast, ecologically constructed roof gardens, known as green roofs or ecoroofs, have a net-positive impact: capturing rainwater to reduce stormwater runoff pollution, covering a large portion of the roof surface to insulate the building and cool the air, and creating habitat for native or migrating species. The design and construction of modern green roofs demands a holistic approach to maximize the benefits to the building and the community. Peter Philippi, a German green roof expert, has said that "a single green roof is like a drop of water in the desert," but it is only in building the most environmentally sensitive individual green roofs that the concrete landscapes of our cities will become functional green infrastructure.

This chapter introduces the basics of ecological green roof design and construction, and outlines the spectrum used to organize the following case studies of exceptional international ecological green roof projects.

Green roof definition, components, and function

At its most basic, a green roof is an engineered roofing system that allows vegetation to grow on top of buildings while protecting the integrity of the underlying structure. The specific materials may vary by project, but every green roof has the same basic components. For the green roof to function properly, it must have a waterproofing membrane, a root barrier, a drainage and water retention layer, growing medium, and plants. The ecological performance of an individual green roof is inextricably tied to the proper function of its components. A healthy, well-established green roof most effectively retains water, cools the air, and insulates the building. An understanding of the materials that make up a green roof is essential to understanding green roof design and ecological function. (Detailed information about spe-

cific materials used to build a green roof can be found in Appendix I.)

Green roofs are generally categorized as *extensive* and *intensive*. *Extensive* green roofs—also known as ecoroofs or roof meadows—are lightweight, low-maintenance, and usually inaccessible. They are often planted with drought-resistant species (sedums, sempervivums, delisperma) that require only 2-6 inches of lightweight substrate (weighing 15-30 pounds per square foot) and subsist on rainwater. Roof meadows generally cover the vast majority of a particular roof, and are generally not built for human occupancy. *Intensive* green roofs—functionally and aesthetically similar to traditional roof gardens—are so-called because of their intensive maintenance requirements. Intensive green roofs are often installed as outdoor amenity space, and buildings featuring them must be able to bear extra weight for human occupancy and more elaborate plantings. Intensive green roofs usually have deeper soil and can accommodate a wide range of plants, edibles, shrubs, and even trees, which in turn require regular maintenance and irrigation. An extensive green roof most simply and cost-effectively meets the goals of ecological design, because of its limited irrigation needs and self-sustaining plantings. An intensive roof can meet the same goals, but issues of irrigation, fertilization, pest control, and plant selection must be carefully considered.

Categorization spectrum

The green roofs showcased in the building case studies section represent the many possibilities of ecological design and construction. The case studies were selected from over 200 submissions from 13 countries. They were selected for exceptional design on the building scale as well as their function as ecological infrastructure within a larger geographic context. The majority of the roofs are low-maintenance extensive designs. Special attention was paid to selecting a wide variety of geographic locations, building types, and innovative designs. The narratives and photos highlight the unique characteristics of each green roof, particularly as the design relates to the ecological function. The technical design details for each case study are listed in Appendix I.

The green roof building case studies are organized along a spectrum based on the three major factors considered in sustainable design— economy, social value, and ecology. While these

themes frame the spectrum, other trends also develop. As the case studies move from economic (and aesthetic) to social to environment, the designs grow in scale, from private development (apartments and single family dwellings), to institutions (museums and schools), municipal infrastructure (public buildings and green neighborhoods), industrial ecology (treatment and industrial plants), ecosystem (animal habitat and plant restoration). (See the photographs at left for a visual representation of the spectrum.) Furthermore, with the increase in scale, the impact of the cumulative green roof benefits widens, beginning with private aesthetic or energy benefits, and moving to communal or neighborhood appreciation, municipal application, regional impact, and ecosystem restoration.

It is assumed that each project considered the economic, social, and environmental impact of its design, but the position of each case study along the spectrum is based on its propensity towards one of the principles. For example, the Mexican CICEANA green roof, which focuses on habitat restoration, is placed toward the end of the spectrum with other environmental projects, although it also has a social aspect in its agricultural plots. The Primary and Secondary School roof in Unterensingen, Germany, includes the economic benefit of solar panels incorporated into the green roof, but it is primarily an educational tool, and thus emphasizes social benefits. The spectrum is presented as a framework to examine green roof design and construction in light of ecological considerations, yet does not preclude other ways of thinking about the projects.

The spectrum also demonstrates how individual green roofs, even those with varied functions, can collectively serve as an ecological infrastructure in a city. A city or region could support many different green roofs for different functions, or invest in green roofs for a single primary purpose, like stormwater management, and still reap a multiplicity of benefits. The seven municipal case studies following the building case studies present some of the possibilities for municipal investment in a green roof infrastructure.

Individually, a green roof represents a shift toward a more sustainable approach to landscape design and urban architecture. Collectively, a network of green roofs becomes part of a city's ecological landscape, connecting parks, community gardens, urban forests, and the built environment.

01 | Glen Patterson's Garden on the Escala

LOCATION
Vancouver, Canada

DATE
2003

CLIENT
Glen Patterson

BUILDING OWNER
Aspac Developments Ltd.

ARCHITECT
James Cheng, K.M. Cheng Architects Inc.

LANDSCAPE ARCHITECT
Jim Nakano, Nakano Landscape Design

GENERAL BUILDING CONTRACTOR
Scott Construction Ltd.

ROOF GARDEN DESIGNERS
Sabina Hill Design Horticulturalist
Paul Tascheraux, Stonecraft Division,
Aquapro Ltd.

SOIL SPECIALIST
Laura Principe, West Creek Farms

GREEN ROOF SIZE
2,000 s.f.

SOIL DEPTH
12-24 in.

When lifetime gardener Glen Patterson moved from his suburban home to a condo in downtown Vancouver's harbor district, he refused to leave his award-winning garden behind. Instead, he brought his backyard along with him, complete with waterfalls, Koi fish, and over 200 species of plants.

Patterson's unit in the new 29-story Escala tower is adjacent to a third-floor roof. When the building was still in its planning phase, he convinced the architect, developer, and engineer to redesign the roof to support his garden. The roof's structure was reinforced, increasing its load capacity from 100 to 250 pounds per square foot.

Preparing the inhabitants of Patterson's backyard for transplant to the roof was more complicated. Two years before the move, workers dug trees up from the backyard and cut the roots back, forming 3-foot diameter balls. This enabled the trees to adapt to the compact environment of a roof. Shortly before the move, the root balls were wrapped tightly in plastic strapping, borrowing a bonsai technique to restrain root growth.

The final garden overlooks the harbor, and its design mirrors the natural landscape of mountains, sea, and forest. Artificial rocks form pathways along a stream, which flows over a waterfall and into a fish pond. Gnarled Japanese maples, one almost 100 years old, flourish among the ginkgos, dwarf rhododendrons, and mountain hemlock. Nearly two years after the move, all the trees and plants are healthy. Despite the heavy root pruning and binding, the trees have adapted to the roof garden environment, and the garden requires almost no maintenance.

COURTESY OF RANDALL SHARP, SHARP & DIAMOND LANDSCAPE ARCHITECTURE AND PLANNING

02 | Casa Bauträger

LOCATION
Linz, Austria

DATE
1996

CLIENT
Casa Bauträger GmbH

ARCHITECT
Ortner & Ortner, Vienna
Zellinger-Landrichtinger-Schrenk

LANDSCAPE ARCHITECT
Thomas Pree, Halbartschlager & Pree
Dachbegrünungs GmbH

ROOFING CONTRACTOR
Walter Ploberger

GREEN ROOF SYSTEM MANUFACTURER
Optigrün International AG

GREEN ROOF SIZE
4,295 s.f.

SOIL DEPTH
15.75 in.

The elaborate Casa Bauträger garden is both a private paradise for its owners and a response to one of the most stringent municipal green roof codes in the world. The Austrian city of Linz requires green roofs on all new residential and commercial buildings with rooftops larger than 1000 square feet (100 square meters). As an incentive, the city offers a reimbursement of up to thirty percent of construction expenses. Instead of responding to this municipal mandate with a simple sedum cover, Herr Bauträger chose to create a lavish retreat around his penthouse apartment.

The intensive garden is on one of the city's tallest buildings, and is accessible only by a private staircase. A cluster of trees and flowers shades the glass-walled living room, and creeping plants insulate the apartment from above. An herb garden flourishes next to the kitchen window. A path leads to a small pond, complete with reeds and lily pads. The dining patio offers views of the city, while strategically-placed rose and raspberry bushes obscure views of the nearby industrial area.

Linz began to mandate green roof development to improve the increasing air pollution caused by industrialization and loss of green space to development. The Bauträger garden not only offers a respite from the surrounding city, but is part of the four million square foot (400,000 square meter) green roof network that benefits the entire urban environment of Linz.

COURTESY OF EDITH ALMHOFER

03 | KPMG

LOCATION	LANDSCAPE ARCHITECT	GREEN ROOF SIZE
Düsseldorf, Germany	Ulrich Zens	44,100 s.f.
DATE	ROOF CONTRACTOR	SOIL DEPTH
2003	Ralf Cremers	12-17 in.
CLIENT	GREEN ROOF CONSULTANT	
KPMG	Marcin Gasiorowski	
ARCHITECT		
Eckhard Gardenier		

As part of a larger building renovation, KPMG in Düsseldorf created a wetland on the roof of its parking garage, an attractive enhancement that also solved stagnation problems in the existing ornamental pond. The German headquarters of the Swiss cooperative, which provides professional services to businesses around the world, is a multi-use building, whose underground garage is covered with a green roof courtyard and pool. When KPMG initiated a renovation to add a conference hall to the site, the company decided to redesign the pond, which often became choked with algae fed by the high phosphate levels in the tap water used to fill it. The new design utilizes the natural cleansing properties

of "multifunctional wetland plants," a patented mix of wetland plants, to filter rainwater for the pond, and expands the landscape into a naturally purifying wetland visible from the new conference hall.

The integrated system has four components: a well, a landscaped area, an artificial swamp and creek, and the original pond. The well collects rainwater, which is pumped out to irrigate the landscaped area outside the conference center. Water bubbles out of the well, into a creek that flows to the swamp, and then drains into the pond. The waterways are lined with volcanic rock and theolite, which further cleans the

water and anchors the plants—species chosen either for their appearance or to purify the water. Workers can enjoy the environment on a wooden deck surrounding the area.

The wetland roof has improved the microclimate around the complex and provides a haven for insects and birds. KPMG has saved money through reduced stormwater fees and energy costs in the underground car park. Ecologically designed green roofs with water features have the potential to revolutionize urban landscape design, fusing the aesthetic amenities of water and plants, the energy savings of traditional green roofs, and the benefits of natural water purification.

COURTESY OF JACEK PLAZA

04 | Heinz 57 Center/Gimbel's Building Rehabilitation

LOCATION
Pittsburgh, Pennsylvania

DATE
2001

CLIENT
623 Smithfield Associates

ARCHITECT
Burt Hill Kosar Rittlemann Associates

LANDSCAPE ARCHITECT
Burt Hill Kosar Rittlemann Associates

GREEN ROOF DESIGNER
Roofscapes Inc.

GENERAL CONTRACTOR
Deklewa Construction Services

GREEN ROOF INSTALLER
Lichtenfels Nursery

PLANT SUPPLIERS
Emory Knoll Farms
Esbenshades Nursery

ROOFING CONTRACTOR
Burns & Scalo Roofing

ROOFING PROVIDER
Carlisle Syntec Inc.

SYSTEM MANUFACTURER
Roofscapes Inc.

GREEN ROOF SIZE
12,000 s.f.

SOIL DEPTH
5 in.

The installation of a green roof on the fourteenth floor of the Heinz 57 Center was an important aspect of the building's transformation from an early twentieth century Gimbel's department store into a modern multi-use structure. The long-abandoned building on the fringes of Pittsburgh's historic downtown underwent a major renovation begun in 1998 that included a soaring 50-foot-diameter octagonal atrium, and a green roof on the 30-foot-wide terrace.

The terrace features a variety of flowering plants, from sedums to wildflowers, that bloom through-out the year. Its integrated decks and paved patio areas can be used for outdoor meetings and gatherings, allowing employees to enjoy the thriving greenery.

The patio is built on roof decks of high-density plastic lumber and employs a paving technique that allows water to flow freely between the paved and vegetated areas. The roof's high walls create a flexible microclimate that enabled designers to specify a wide range of drought-tolerant plants. The nursery provided a two-year maintenance agreement to ensure the health of plantings until the roof became established, at which point it became virtually maintenance-free.

This green roof is visible from most of the major downtown office towers, and serves as a model to encourage green roof development that could reduce downtown Pittsburgh's heat island effect. The developers have brought the roof to the attention of the mayor, in hopes of encouraging government and private sector interest in widespread green roof implementation.

COURTESY OF BURT HILL KOSAR RITTLEMAN ASSOCIATES

COURTESY OF ROOFSCAPES INC

HEINZ 57 CENTER/GIMBEL'S BUILDING REHABILITATION

05 | North German State Clearing Bank (NORD LB0)

LOCATION
Hannover, Germany

DATE
2002

CLIENT
DEMURO Grundstücksverwaltung
mbH &Co

ARCHITECT
Behnisch, Behnisch & Partners

SYSTEM MANUFACTURER
ZinCo GmbH

GREEN ROOF SIZE
84,584 s.f.

SOIL DEPTH
4.5-8 in.

The striking angles, cantilevers, glass façade, and terraced green roofs of the Norddeutsch Landesbank building in Hannover were designed as an exaggerated likeness of the surrounding cityscape. The German bank's new energy-efficient green building has become a social and commercial hub in a large corporate park of 1,500 employees.

The building's stepped architecture allows for terraced green roofs on the first 18 floors. The landscaping is not meant to be a separate de-sign element, but to emphasize the structure's sparse, modern lines. Most of the terraces are planted with a single variety of closely controlled sedums to accentuate the linear style. Because the façade is clear glass, the palette of the exterior roof plantings impacts the interior aesthetic, so the roof terrace adjacent to the staff restaurant is planted instead with bright seasonal flowers, including poppies, oxeye daises, and dahlias. Accessible terraces feature decks of red cedar, reclaimed wood, or stone lined with potted plants.

The multi-level green roof is also important for climate control in accordance with Germany's low-energy building regulations. Along with a rainwater pond in the central courtyard, the roof greenery creates a pleasant microclimate in the building. The structure was designed to comply with the full range of green building laws. It receives eighty percent natural daylight year-round, and utilizes geothermal chilling, fuel cell heat, external sunshades, window ventilation, and a double skin façade.

COURTESY OF CHRISTIAN KANDZIA

06 | A&C Systems

LOCATION
Hoegaarden, Belgium

DATE
2001

CLIENT
A&C Systems

ARCHITECT
Mark Depreeuw

LANDSCAPE ARCHITECT
Geert Meysmans

ENGINEER
Geo Van Nieuwenhove

CONTRACTOR
Moeskops Bouwbedrijf

SYSTEM MANUFACTURER
FLORADAK

GREEN ROOF SIZE
3,229.2 s.f.

SOIL DEPTH
4.3 in.

The A&C Systems building in the Belgian town of Hoegaarden was designed to respect both ecology and human psychology. The brightly colored sedum roof was an essential part of the architect's vision for a building that is both energy efficient and inviting to the community and workers—a model for what he calls "bio-ecological architecture."

Despite its elaborate, colorful design, the roof is a simple and inexpensive extensive green roof system. The drought-resistant sedum varieties grow in a lightweight, shallow substrate and require little maintenance.

The company agreed that the savings on insulation and energy use, plus increased roof membrane lifespan, would pay back the initial expense of the green roof. The temperature reduction realized by the roof has eliminated the need for air conditioning in the building during the summer. Although there are no stormwater regulations in Hoegaarden, rooftop rainwater collectors capture excess rain runoff, which is integrated into the plumbing system. The building boasts energy efficiency beyond the green roof; it returns power to the electrical grid from its windmill and solar panels.

The 3,230 square foot green roof has become a source of community and employee pride. The A&C Systems building demonstrates many of the economic and social benefits of transforming traditional office buildings into distinctive ecological infrastructure.

COURTESY OF ARCHITECTEN ATELIER MARK DEPREEUW

07 | Sechelt Justice Services Centre

LOCATION
Sechelt, Canada

DATE
2003

CLIENT
District of Sechelt

ARCHITECT
Johnston Davidson Architecture

LANDSCAPE ARCHITECT
Sharp & Diamond Landscape Architecture & Planning

CONTRACTOR
Vanbots Construction Corporation

PLANT SUPPLIER
Peels Nursery Canada

SYSTEM MANUFACTURER
Soprema

GREEN ROOF SIZE
5,000 s.f.

SOIL DEPTH
3 in.

The Sechelt Justice Services Centre, one block from the Pacific Ocean in British Columbia, is a unique example of low-impact, site-specific development. The building's layout and location take advantage of sunlight and prevailing winds for optimum energy efficiency. The green roof mimics local ecology in order to replace a meadow lost to the development.

Many of the sedums and grasses on the roof also grow in the adjacent coastal meadow. These hardy native species are planted alongside appropriate non-native plants in a growing medium made up of local materials. Black pumice from a volcano near Quesnel, British Columbia, is mixed with recycled hemlock and fir bark from the local lumber industry, mushroom compost, and landscape green waste. The high percentage of organic matter provides nutrients and microbial activity to enhance rapid plant growth.

The green and blue grasses on the roof are complemented by green, red, and silver sedums, and go dormant in the summer, turning shades of gold and brown. The plants turn green again in the mild northwest winter, and in the spring, the sedum flowers attract insects and songbirds to the roof. Though the roof is inaccessible, employees and visitors enjoy its bright ever-changing colors from above and below.

COURTESY OF THE DISTRICT OF SECHELT

COURTESY OF RANDALL SHARP, SHARP & DIAMOND LANDSCAPE ARCHITECTURE AND PLANNING

08 | Hill House

LOCATION
La Honda, California

DATE
1979

CLIENT
George & Adele Norton

ARCHITECT
Jersey Devil Design/Build

LANDSCAPE ARCHITECT
Jersey Devil Design/Build

ENGINEER
Glenn Nelson

GENERAL CONTRACTOR
Jersey Devil Design/Build

GREEN ROOF SIZE
2,500 s.f.

SOIL DEPTH
8 in.

To comply with the local planning commission's mandate for "quietness, repose, and unobtrusiveness" in new construction, Hill House was designed to function in harmony with the environment and blend into the natural terrain. The meticulously sculptural earth house is barely visible in the ridgeline of northern California's "scenic corridor."

The house cuts into the crest of the hill, sheltered from Pacific storms but open to sunlight. The curve of the structure creates a protected outdoor courtyard. The hillside had to be reshaped to accommodate the concrete shell of the building. Lichen-covered boulders unearthed during the digging were incorporated into the retaining wall, replacing the reinforced concrete in places, and further merging building and landscape. Climate control in the house was designed to be responsive to the site's variable microclimate. Cool air cycles through a passive solar heating system, and warm air naturally vents back into the house as it rises.

The sod roof, earth berms, and the fieldstone used in the veneer and retaining walls enable the building to virtually disappear into the landscape, while reducing energy needs and making the house resistant to wind, fire, and earthquakes. Planted with winter rye, native plants, and wildflowers, the green roof slopes down to join the interior courtyard, which is landscaped with honeysuckle vines and an oak tree. The interplay of form and materials creates a unique home seamlessly integrated into its natural surroundings.

COURTESY OF JERSEY DEVIL DESIGN/BUILD

09 | Nine Houses

LOCATION
Dietikon, Switzerland

DATE
1993

CLIENT
Greuter AG Hochfelden

ARCHITECT
Peter Vetsch

ENGINEER
Peter Vetsch

GREEN ROOF SIZE
37,674 s.f.

SOIL DEPTH
2.3-6.5 ft.

Locals joke that the Nine Houses development is home to prehistoric cave people or Tolkien's mythical hobbits, but the luxury underground homes represent a transformative new vision for architecture and housing. The houses, built into and under a hill in the Swiss suburb of Dietikon, are ecologically constructed and energy efficient.

The turf-roofed development offers a sharp contrast to the conventional one-family units in the rest of the neighborhood. The houses cluster around a small lake in an east-facing horseshoe, and are united under 60 tons of earth. One to two meters of earth covers each house, laid atop rounded roofs made of about 6 to 7 inches (15-20 centimeters) of concrete, tarred paper, and an insulating foam of recycled glass. The roof construction creates an insulation layer that eliminates the need for air conditioning or significant heating. When heat is required, it comes from the earth—the geothermal system pumps heat up through 300-foot (100 meter) deep underground ducts. The houses use half the energy of normal homes.

The units have spiral staircases, fanlight windows, and are dry and bright year-round. The green roofs are planted with native grasses, creating a unified landscape and a compelling model for environmentally sound suburban housing.

COURTESY OF PETER VETSCH

10 | Life Expression Chiropractic Center

LOCATION
Sugar Loaf, Pennsylvania

DATE
2001

CLIENT
Life Expression Wellness Center

ARCHITECT
Van der Ryn Architects

LANDSCAPE ARCHITECT
Roofscapes Inc.

ENGINEER
Weir Andrewson Associates

GREEN ROOF DESIGNER
Roofscapes Inc.

CONTRACTOR
Houck Services

PLANT SUPPLIERS
Emory Knoll Farms
Esbenshades Nursery

SYSTEM MANUFACTURERS
Roofscapes Inc.
Sarnafil

GREEN ROOF SIZE
6,000 s.f.

SOIL DEPTH
5 in.

The green roof atop the Life Expression Chiropractic Center is a visual expression of the Center's holistic approach to well-being. The dramatically sloped 6,000 square foot green roof is part of an environmentally sensitive building design that also features passive solar heat, natural daylighting, and paints low in volatile organic compounds. Functionally, the roof regulates interior temperatures and controls runoff. Aesthetically, rooftop vegetation enables the building to blend into the surrounding rural Appalachian valley.

The steep 30-degree pitch of the roof and exposure to the region's intense winds presented engineering challenges during construction. Roof battens, slope restraint panels, and reinforcing mesh were used at various points to stabilize the plantings. To avoid erosion until the plants were established, the roof was covered with a photodegradable mesh wind blanket, which has since been covered by the mature plants.

The green roof reduces the rate and quantity of rainwater runoff. Rain sheets off the roof through gaps in the fascia, and drips down the tendrils of overhanging plants. The cascading vegetation creates a curtain effect along the eaves. Soothing details like this support the healing mission of the Center and allow the building to blend seamlessly into the adjacent landscape.

COURTESY OF ROOFSCAPES INC

11 | Arbroath Abbey Visitor Centre

LOCATION
Arbroath, Scotland

DATE
2001

CLIENT
Angus Council

ARCHITECT
Simpson & Brown Architects

LANDSCAPE ARCHITECT
John Richards Landscape Architect

ENGINEER
Wren & Bell

MAIN CONTRACTOR
Clachan Construction

ROOF SUPPLIER
Erisco Bauder

ROOFING CONTRACTOR
Advanced Roofing Services

SYSTEM MANUFACTURER
Erisco Bauder

GREEN ROOF SIZE
1,137 s.f.

SOIL DEPTH
1 in.

Designing a contemporary visitor's center appropriate for the imposing stone ruins of the Arbroath Abbey in Scotland posed a significant design challenge. The final structure is a layered, horizontal form, a strong visual contrast to the vertical lines of the gothic abbey. The center, with its low-impact modern design, houses a retail shop, education center, and gallery space. The building was constructed of natural materials—stone, timber, glass, and a multi-colored sedum-covered green roof intended to mirror the adjacent sloping public green.

Each of the center's several levels is brighter and more open than the one below, culminating in a gallery suspended over the rear graveyard wall, which offers a complete view of the ruins beyond. The elegant trusses combine traditional and modern materials. The structural joints are left exposed, allowing visitors to see how the building was constructed. Timbers for the segmented upper roof are naturally curved, eliminating the need for glued and laminated beams.

The roof was chosen for both practical and aesthetic reasons: a green roof was determined to be the most appropriate, low maintenance, and ecologically sound method of covering such a large, low-pitched roof. The mixed sedums are rooted in a mineral wool blanket, which stores water and helps the vegetation survive during dry periods. The plantings' colors change with the season, adding variation and texture to the unchanging historic stone abbey.

COURTESY OF KEITH HUNTER

ARBROATH ABBEY VISITOR CENTRE

12 | Mashantucket Pequot Museum and Research Center

LOCATION
Mashantucket, Connecticut

DATE
1997

CLIENT
Mashantucket Pequot Tribal Nation

ARCHITECT
Polshek Partnership Architects

LANDSCAPE ARCHITECT
Office of Dan Kiley

ENGINEER
Ore Arup & Partners, New York, New York

SYSTEM MANUFACTURER
American Hydrotech

GREEN ROOF SIZE
52,000 s.f.

SOIL DEPTH
9-36 in.

The Mashantucket Pequot Museum and Research Center celebrates the history and survival of the Pequot nation. The museum's unique design, which uses organic forms and green space, visually expresses the Pequots' connection to the surrounding land, agriculture, and water.

The building's three integrated shapes—a circle, an organic landform, and a linear bar—contain the institution's three primary elements—gathering space, museum, and research center. The museum's roof is a two-tiered terrace, land-scaped with native plants. The carefully-controlled green roof was designed to emulate a natural clearing in the woods, the kind that protected and provided for generations of Pequots. The rooftop's orderly indigenous plantings contrast with the undisturbed forest and swamp beyond, heightening the impact of the view.

From inside the museum and the attached circular gathering space, glass walls offer views of forest and sky. The linear atrium houses archives, a theater, and a library, where visitors can learn more about Pequot history. The three horizontal shapes are juxtaposed with an accessible tower, which symbolizes the cultural rebirth of the Mashantucket Pequots and provides views of the land that shaped their culture and history.

The green roof lessens the building's impact on the adjacent wetland, while visually blurring the distinction between building and landscape. Visitors to the fully accessible roof can explore the Mashantucket Pequot's harmonious relationship with the land.

COURTESY OF JEFF GOLDBERG/ESTO

13 | Schiphol Plaza

LOCATION
The Hague, The Netherlands

DATE
1994

CLIENT
Amsterdam Airport Schiphol

ARCHITECT
Benthem Crouwel NACA Amsterdam/
Den Haag

LANDSCAPE ARCHITECT
Andriaan Geuze, West 8

ENGINEER
Bureau de Weger Rotterdam

CONTRACTOR
KLS 2000 Schiphol

SYSTEM MANUFACTURER
Xeroflor

GREEN ROOF SIZE
87,188.4 s.f.

SOIL DEPTH
1.3 in.

The Schiphol Plaza at Amsterdam's airport, opened in 1995, is in many ways an ideal model of airport design. After almost thirty years of building, extension, and renovation projects, the final floor plan enables all pedestrian traffic to and from the terminals to converge in a single building. Transferring and changing flights is possible under one sedum-planted roof.

The 26,500 square foot extensive green roof is part of the architectural goal of merging in-door and outdoor spaces in a major transportation hub. The huge Schiphol Plaza hall provides a connection point for all the airport's terminals and the Dutch Rail train station. The elegant glass-walled space resembles an outdoor town square, and houses more than forty stores, kiosks, and restaurants. Breezes enter the hall through gaps under the roof, providing most of the necessary climate control.

From above, the roof is reminiscent of the surrounding flat plains. The sedums change color with the season, greeting travelers with green in winter, pink in spring, and vibrant red in summer. In accordance with the airport's stringent environmental and landscape requirements, the green roof lowers energy costs and absorbs rainwater runoff. Schiphol's rooftop—the only known airport green roof—embodies the fusion of aesthetics and ecology in green roof design.

COURTESY OF XEROFLOR

COURTESY OF BENTHEM CROUWEL

COURTESY OF BENTHEM CROUWEL

COURTESY OF XEROFLOR

14 | Primary and Secondary School

LOCATION
Unterensingen, Germany

DATE
2002

CLIENT
Spitzmaus GbR

ARCHITECT
Ulrich Kahl

GREEN ROOF CONTRACTOR
Falk Ruoff Aichtal-Aich

SYSTEM MANUFACTURER
ZinCo GmbH

SOLAR MANUFACTURER
Solar-Fabrik AG

GREEN ROOF SIZE
15,000 s.f.

SOIL DEPTH
3 in.

Cutting-edge green technologies are transforming the classroom experience of students in Unterensingen, a town near Stuttgart. The town mandated "ecological and economical use" for the renovation of the Primary and Secondary School roof. The result is an extensive green roof with integrated photovoltaic panels for solar energy collection—which doubles as a student laboratory.

Students monitor the energy generated by the solar panels, and estimate the volume of carbon dioxide that would have been produced if it had been generated by fossil fuels. Evaluating these figures and measuring green roof tempera-

ture and water levels are part of the school's science curriculum. A display in the main entrance lets students check current and cumulative power output, and a viewing terrace allows them to experience the green roof first-hand.

Solar panels and green roofs may be the synergistic green building combination of the future. German research demonstrates that solar panels function optimally at temperatures under 77°F; a conventional roof gets much warmer during the summer. Green roofs lower ambient roof temperature, enabling panels to operate more efficiently. Solar panels incorporated into green roofs like that in Unterensingen have been

shown to generate fifteen percent more energy than those on asphalt- or gravel-topped roofs.

Solar energy incentives in Germany often mandate that the buyback price of renewable energy entering the power grid be more than twice that of normal electricity. The utility company pays Unterensingen 55cents/KwH for its solar power, compared to 20cents/KwH for standard electricity. The school expects the roof to pay for itself in less than 10 years. The Primary and Secondary School solar green roof is stimulating the sustainable energy market, while giving children firsthand experience of the benefits of green technologies.

COURTESY OF ZINCO GMBH

15 | Beddington Zero Emission Development (BedZED)

LOCATION
London, England

DATE
2002

CLIENT
Peabody Trust

ARCHITECTS
Bill Dunster Zedfactory Ltd

LANDSCAPE ARCHITECT
Andrew Grant Associates

ENGINEER
Ellis and Moore

SUSTAINABILITY ADVISOR
BioRegional

SYSTEM MANUFACTURER
RAM Roof Garden Consultancy Ltd

GREEN ROOF SIZE
333,519 s.f.

SOIL DEPTH
1-11.8 in.

On the site of a former sewage treatment plant in a densely developed part of London, a new model for affordable housing has emerged. City planners estimate that London will need 3.8 million new homes by 2016—a nineteen percent increase since 1991. Even if new housing is built at current density levels, it would cover a land area larger than Greater London. Beddington Zero Emission Development, or BedZED, demonstrates how high-occupancy, high-quality housing can be built without significantly impacting the environment. BedZED, which can house one hundred families, is the first large-scale "carbon neutral" community in the world, offsetting its carbon dioxide emissions with renewable energy features.

The development makes use of simple, proven, sustainable methods. Each unit integrates residential and office space. The workspace rooftops become garden spaces for the housing units above, so that every home has a private garden, even at building densities that would normally only allow for a balcony. The extensive sedum mats cover-ing the topmost roof surfaces reduce dependence on the sewer system by absorbing runoff.

BedZED homes maximize natural lighting with passive solar design. All wood used is certified sustainable. Almost all materials are local, sourced from within a 35-mile radius to reinforce local identity and reduce transport energy and pollution. The units are designed to accept photovoltaic panels; if installed, these solar panels could provide power for forty electric cars, further reducing the need for fossil fuels.

COURTESY OF BILL DUNSTER ARCHITECTS

16 | EcoHouse

LOCATION
St. Petersburg, Russia

DATE
2002

CLIENT
Cooperative Apartment Block at Pulkovskaya 9/2

LANDSCAPE ARCHITECT
Alla Sokol

GREEN ROOF SIZE
18,300 s.f.

SOIL DEPTH
1.57–3.15 in.

In a country where waiting on bread lines was once a daily ritual and food security has been a concern for almost a century, rooftop food production seems a logical next step in urban development. EcoHouse, in the Moskovsky district of St. Petersburg, has transformed a 1960s high-rise apartment block into a flourishing sustainable community with an agricultural roof garden. Responding to the difficult social, economic, and environmental conditions of the new Russia, EcoHouse serves as a replicable model for converting Soviet-era housing blocks into self-contained, sustainable communities.

EcoHouse draws on a largely untapped labor resource—sixty percent of the building's five hundred residents are pensioners, and many others are unemployed. The program has taught participants how to develop small businesses; they sell both compost and produce from the rooftop. The process partially closes the "resource loop," keeping the community's products, cash, and labor within the building, and decreasing reliance on sometimes scarce external resources.

The rooftop garden is the heart of EcoHouse, producing food, improving air quality, creating green space, and fostering camaraderie. Fruits and vegetables grow in plastic grocery flats and greenhouses in a compost substrate. Shallow beds of grass cover the remaining asphalt, and flowers in planters fashioned from old tires decorate open areas. The rooftop, created at minimal cost with limited resources, provides food, employment, and a new sense of community for the building's residents.

COURTESY OF ALLA SOKOL

17 | St. Luke's Science Center Healing Garden

LOCATION
Tokyo, Japan

DATE
1992

CLIENT
St. Luke's International Hospital

ARCHITECT
Nikken Sekkei

LANDSCAPE ARCHITECT
Nikken Sekkei

ENGINEER
Nikken Sekkei

GENERAL CONTRACTOR
Shimizu Corporation

LANDSCAPE CONTRACTOR
Uchiyama Landscape Construction Co.

GREEN ROOF SIZE
15,500 s.f.

SOIL DEPTH
19.7-27.6 in.

Studies have shown that green open spaces have tangible benefits to human health and well-being, speeding recovery of hospital patients and calming the very ill. Urban hospitals are often unable to offer these benefits to their patients, but St. Luke's Science Center found a solution by creating a healing garden on its open rooftop. Patients, visitors, and workers can enjoy views of Tokyo as they stroll the green roof's meandering garden paths through lush greenery.

The roof garden is open to both hospital patients and the public for rehabilitation, exercise, and socializing. The popular children's playground areas are available to young patients and visitors. Tours are offered to local civic groups and schools interested in learning about roof gardens.

Part of a larger renovation to improve the hospital's aging facilities, the rooftop garden was designed as a peaceful refuge where patients could enjoy fresh air and sunlight to enhance their healing. The space serves to remind visitors of the hospital's four inter-related values: faith, medicine, community, and ecology.

COURTESY OF ST. LUKE'S INTERNATIONAL HOSPITAL

18 | Chicago City Hall

LOCATION
Chicago, Illinois

DATE
2001

CLIENT
City of Chicago

ARCHITECT
William McDonough + Partners

LANDSCAPE ARCHITECT
Conservation Design Forum

ENGINEER
Halvorsen & Kaye Associated
PC Katrakis & Associates

ROOFING CONTRACTOR
Bennet & Brosseau Roofing

PLANT SUPPLIERS
Bluebird Nursery
Intrinsic Perennial Garden
Midwest Groundcovers
The Natural Garden
Teerling Nursery

ENVIRONMENTAL ENGINEER
Atelier Dreiseitl

SYSTEM MANUFACTURER
Roofscapes Inc.

MEMBRANE MANUFACTURER
Sarnafil

GREEN ROOF SIZE
22,000 s.f.

SOIL DEPTH
4 in., 6 in., 18 in.

As part of its effort to become the "greenest city in America," Chicago now claims more green roofs than any other U.S. city. The elaborate green roof on the century-old City Hall was among the first, and is now an important research and demonstration site for studies on the benefits of green roofs, comparisons of green roof technology, and the survivability of both native and non-native plant species.

The sunburst pattern of the plantings includes such species as the broad-armed prairie crabapple and orange-flowered butterfly weed. Over 150 varieties of trees, vines, grasses, and shrubs are the subjects of ongoing experiments. Plants are organized into bands of different colors, which change as the season progresses. These bands are not merely aesthetic, but allow the same plant material to be tried in various soil depths, slopes, and drainage patterns. Although the roof is not accessible to the public, it is visible from more than thirty tall buildings in the city center.

The black tar roof of the adjacent Cook County Administration Building serves as a control for comparative studies. Tests have shown that the average temperature on the City Hall green roof is 78°F cooler than on the Administration Building's tar roof. Chicago's City Hall green roof continues to provide the essential data that can ultimately support government incentives and investment in green roofs as a way to mitigate the urban heat island effect.

COURTESY OF CONSERVATION DESIGN FORUM

19 | ACROS Fukuoka

LOCATION
Fukuoka, Japan

DATE
1995

CLIENT
Dai-Ichi Mutual Life Mitsui Real Estate

ARCHITECT
Emilio Ambasz
Nihon Sekkei Tekenaka Corporation

LANDSCAPE ARCHITECT
Nihon Sekkei Takenaka Corporation

ENGINEER
Nihon Sekkei Takenaka Corporation
Consultant: Plantago Corporation

SYSTEM MANUFACTURER
Katamura Tekko Company

GREEN ROOF SIZE
100,000 s.f.

SOIL DEPTH
12-24 in.

The Asian Crossroads Over the Sea (ACROS) building in crowded Fukuoka offsets the impact of its development on the adjacent Tenjin Central Park with a series of elaborate stepped green roofs. The project is an articulate fusion of public and private space, which more than doubles the size of the park while creating over one million square feet of multipurpose space, including a museum, a theater, shops, and offices.

The city-owned land was the last large undeveloped plot in central Fukuoka and is next to the only park in the area. The city chose to develop the site in a joint venture with private enterprise. The goal was to create new public land equal to that lost to the development. The north face of the building is traditional, with a formal entrance onto one of the most prestigious blocks in the city. On the south side, 15 vegetated terraces climb the full height of the building. Each floor has landscaped gardens for meditation, relaxation, and escape from the congestion of the city below. The top terrace is a grand belvedere, with views of the bay of Fukuoka and the surrounding mountains. A series of reflecting pools are connected by water jets, creating a climbing waterfall that masks the ambient noise of the city.

The design reconciled the developer's desire for profitable site use with the public need for open space. The nearby park and the ACROS building are inseparable, demonstrating how a major building complex can coexist with public green space.

COURTESY OF HIROMI WATANABE/EMILIO AMBASZ & ASSOCIATES

20 | Church of Jesus Christ of Latter Day Saints

LOCATION
Salt Lake City, Utah

DATE
2000

CLIENT
Church of Jesus Christ of Latter Day Saints

ARCHITECT
Zimmer Gunsul Frasca Partnership

LANDSCAPE ARCHITECT
Olin Partnership

ENGINEER
KPFF Consulting Engineers

GENERAL CONTRACTOR
Legacy Constructors

ROOFING CONTRACTOR
American Hydrotech

SYSTEM MANUFACTURER
American Hydrotech

LANDSCAPE CONTRACTOR
American Landscape

GREEN ROOF SIZE
174,240 s.f.

SOIL DEPTH
2-48 in.

The Conference Center of the Church of Jesus Christ of Latter Day Saints in Salt Lake City is one of the world's largest religious buildings, accommodating 21,000 congregants in its 1.1 million square feet of interior space. Integrating this massive structure into the immediate environment and the greater landscape of the Wasatch and Oquirrh mountain ranges was a challenge for the church designers. A 65-foot change in elevation along the site allowed the architects to submerge parts of the structure into the landscape, connecting different levels with an extensive system of exterior stairs, terraces, green roof gardens, and fountains.

The roof's 4-acre garden and 3-acre meadow were designed with a lightweight growing medium and careful attention to the structural limitations of the assembly hall. Substrate beneath the meadow is only 2 inches deep, but slopes up gradually to a depth of four feet in areas where trees are planted. More than a thousand volunteers carried flats of plants from the street to the rooftop in bucket-brigade style, hand-planting every grass and perennial in the meadow.

Rooftop vegetation on this scale eliminates the need for the intensive heating and cooling normally required for such a large assembly hall. The green roof controls runoff on the site by absorbing virtually all rainwater. In addition to the ecological benefits, the vast expanse of native grasses, firs, pines, and aspens creates an oasis for meditation and contemplation, and a gathering place for the congregation.

COURTESY OF ECKERT & ECKERT

CHURCH OF JESUS CHRIST OF LATTER DAY SAINTS

21 | Osaka Municipal Central Gymnasium

LOCATION
Osaka, Japan

DATE
1996

CLIENT
Osaka City Government

ARCHITECT
Nikken Sekkei

LANDSCAPE ARCHITECT
Osaka City Government

ENGINEER
Nikken Sekkei

GENERAL CONTRACTORS
Asanuma Corporation
JV, Obayashi Corporation
Nishimatsu Construction Company

GREEN ROOF SIZE
1,101,089 s.f.

SOIL DEPTH
40 in.

Although it is known as the City of Water, in recent years Osaka has suffered from the decline in popularity of the marine transportation that had once spurred the city's economic growth, eroding its surrounding green space. The municipal government has begun to promote urban greening policies to remedy the resulting environmental problems, including poor water quality and increased urban heat island effect. An 82-foot hill at one end of the Yahataya Park demonstrates the city's new commitment to green buildings.

The 360-foot diameter hill covers the rooftop of the Osaka Municipal Central Gymnasium, a proposed venue for the 2008 Olympics. The building is sunk 30 feet into the ground and has an average of 3 feet of soil covering the entire structure. A wall foundation supports the 70,000-ton earthen roof. The earth insulates the arena, while geothermal heat warms it in the winter. A 55-foot opening at the apex of the dome provides natural light and air circulation to further maintain a comfortable interior environment.

With a seating capacity of 10,000 and a roof that is considered an essential element of the city's parks system, this unusual gymnasium is at once a sports center and a public park. Paths lead to the roof's peak and benches there provide magnificent views of the city. The area is planted with flowers and trees so that it appears to merge seamlessly with the rest of Yahataya Park. The city's choice to conserve green space by building its Olympic stadium underground establishes a new standard for green municipal construction.

COURTESY OF YOSHIHARU MATSUMURA

22 | Daimler Chrysler Complex

LOCATION
Berlin, Germany

DATE
1999

CLIENT
Daimler Chrysler

SITE DESIGNER
Renzo Piano

ARCHITECTS
Arata Isozaki
Hans Kollhoff
Lauber + Wohr
Rafael Moneo
Renzo Piano
Richard Rogers

WATER ARCHITECT
Atelier Dreiseitl

LANDSCAPE ARCHITECT
Daniel Roehr, Krüger + Möhrle

STORMWATER ENGINEER
Marco Schmidt

SYSTEM MANUFACTURERS
Colbond Geosynthetics/EnkaDrain
Eggers Soil
ZinCo GmbH

GREEN ROOF SIZE
172,224 s.f.

SOIL DEPTH
4-23.6 in.

Berlin's Potsdamer Platz, where Checkpoint Charlie once stood as a symbol of the divided Germany, is today the heart of the unified country's capital and the location of an innovative renovation. The 680,000 square foot (68,000 square meter) Daimler Chrysler complex dominates the site, with 17 buildings and over 400,000 square feet (40,000 square meter) of green roofs. The roofs are part of a stormwater management plan created to meet city regulations, which mandate that new development sites absorb at least ninety-nine percent of rainwater runoff. In addition to almost 40 green roofs, the site includes rainwater barrels, a grey water system, and an artificial lake.

The building complex was designed as an ecological corridor, bordered by the artificial lake and Berlin's largest park. Each of the 17 buildings has both intensive and extensive green roofs. The intensive gardens were individually tailored to complement the architecture of each building, and meet another city requirement for inclusion of "play spaces" or courtyard areas. The colorful extensive green roofs are planted with patterns of white, yellow, and pink sedums.

Runoff from the green roofs was intended to refresh the lake and keep the water from stagnation. However, the roofs are too effective as a stormwater management tool—they retain sixty to seventy percent of runoff, not allowing enough to reach the lake. Chrysler was forced to activate a mechanical pump and filter to keep the lake water clean. Nonetheless, the complex showcases the integration of imaginative landscape-based stormwater management. Research conducted on runoff and retention will be valuable to future projects of this kind.

COURTESY OF VINCENT MOSCH

23 | Roppongi Hills

LOCATION
Tokyo, Japan

DATE
2003

CLIENT
Roppongi 6 Chrome Redevelopment
Association

ARCHITECTS
Conran & Partners
JPI
KPF
Mori Building Company

LANDSCAPE ARCHITECT
Yohji Saski
Dan Pearson

SYSTEM MANUFACTURER
Green Mass Damper Technology

GREEN ROOF SIZE
143,000 s.f.

SOIL DEPTH
1.17-46.8 in.

Roppongi Hills is an experimental urban development in the heart of Tokyo. The complex was envisioned as a way to revitalize a depressed downtown enclave, not only through economic development, but also by adding much needed green space in a city that suffers from intense urban heat island effect. The Tokyo Metropolitan Government worked with the developer Mori Building Company to design Roppongi Hills as an entertainment district and a lush garden neighborhood in the heart of the city.

The area's skyscrapers maximize available space for housing, offices, and entertainment in the dense center city. Site planners were creative in their use of greenery, knitting together the complex with a network of pathways, gardens, and green roofs. In a city of only fourteen percent green space, Roppongi Hills has an unprecedented twenty-six percent of its land area planted with vegetation.

The Keyakizaka complex rooftop boasts a rice paddy and vegetable plot, while the Sakurazaka roof exhibits public art and street furniture in a garden setting. There is a 43,000 square foot traditional Japanese garden, and most of the residential area is designed in a blended Japanese-British style. Almost all the buildings, including the Asahi Television tower and the Virgin Cinema complex, have green roofs. The variety of landscaping on the Roppongi Hills buildings showcases the potential for inventive green and vertical urban development.

COURTESY OF MORI BUILDING COMPANY

24 | Vastra Hamnen

LOCATION
Malmö, Sweden

DATE
2001

CLIENT
City of Malmö

ARCHITECT
Various

LANDSCAPE ARCHITECT
Various

ENGINEER
Various

SYSTEM MANUFACTURER
Veg Tech AB

GREEN ROOF SIZE
48,000 s.f.

SOIL DEPTH
1.4 in.

In response to a growing population, a deteriorating industrial zone, and an expanding role as a gateway to the Baltic Sea, officials in Vastra Hamnen, the old port neighborhood in the Swedish city of Malmö, envisioned an environmentally-conscious urban revitalization. The city selected architects to design sustainable houses as part of a mini-ecological city, dubbed "Bo01," to be showcased at the European Housing Expo in 2001. Designs had to meet "green area factor" specifications—a system which assigned a numeric value to green features such as climbing ivy, bat nesting boxes, vegetable gardens, and green roofs. An open grassy area was worth

1.0 point, a green roof earned 0.8 points, and a house with a conventional roof received 0 points. Development proposals were required to have an average of 10 points per house, from a possible 35 points in total.

As a result, 16 of the 22 Bo01 houses have green roofs, for a total of 48,000 square feet in the complex. The roofs are part of a green stormwater management network, which also includes courtyards and public parks. Excess rainwater flows into a citywide system of ponds and tree-lined channels that run to the sea.

The houses in the Bo01 expo demonstrate a variety of architectural styles, but all follow a specified color scheme, with pale façades facing the sea and brighter colors in the interior. Houses are closely spaced to make the best use of land space, and buildings on the waterfront are taller to provide wind protection for the rest of the neighborhood. The narrow streets are accessible only to pedestrians and bicyclists. Each house includes waste separation and is powered by a combination of solar, wind, and water energy.

COURTESY OF JOHAN THIBERG

25 | GENO Haus

LOCATION
Stuttgart, Germany

DATE
1990

CLIENT
GENO Haus

LANDSCAPE ARCHITECT
Peter Philippi

GREEN ROOF CONTRACTOR
Dachgarten Baubegrünung GmbH

SYSTEM MANUFACTURER
Optima

GREEN ROOF SIZE
30,000 s.f.

SOIL DEPTH
4-10 in.

In Stuttgart, Germany, modern green roofs have been around for more than thirty years. The tallest building in the GENO Haus complex, which houses the state banking coalition, has had a green roof since 1969. Stuttgart lies in a valley, and has long suffered from pollution and artificial microclimates caused by slow-moving air. GENO Haus was a site for early green roof experimentation to mitigate this problem. The project, built in collaboration with local government and early green roof manufacturers, preceded and helped influence the first municipal greening legislation in Stuttgart.

The original 1969 roof consisted of a modular system made of Styrofoam, which preceded the technological development of drainage and root repellant layers. All of the green roofs in the complex remained watertight for over twenty years, but they were renovated in 1990 due to concerns about a possible breach because there was no root barrier layer. Renovation coordinator Peter Philippi designed a series of modern intensive, semi-intensive, and extensive roofs for the complex, which includes an automatic irrigation system that is

activated automatically when water in the drainage layer gets low. Employees use the roof space for recreation during breaks.

The GENO Haus green roofs are an easily replicable model of low-cost, low-maintenance, functional, and ecological green roof application. The complex represents a visionary move by the municipal government in the early years of green roof technology, and was a critical catalyst of subsequent green roof development.

COURTESY OF PETER PHILIPPI

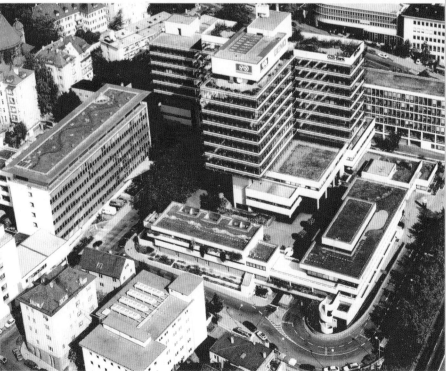

26 | Schachermayer Company

LOCATION Linz, Austria	LANDSCAPE ARCHITECT Halbartschlager & Pree Dachbegrünungs GmbH	GREEN ROOF SIZE 114,829 s.f.
DATE 2000	ENGINEER Gerd Gessner	SOIL DEPTH 4 in.
CLIENT Schachermayer Company	SYSTEM MANUFACTURER Optigrün	
ARCHITECT Gerd Gessner		

The city of Linz has some of the most stringent green roof policies in all of Europe. Since 1989, the Austrian city on the banks of Danube has promoted the development of a citywide green roof infrastructure through both regulations and incentives. The expansion of Linz's steel and chemical industries in the last 20 years accelerated new construction and urbanization. To offset the impact of development, the city instituted a "green plan" in high-density areas. Modifica-

tions to the building code require that all new buildings and underground garages include eighty percent green roof coverage, and specify substrate depth.

The 12-building Schachermayer Company complex demonstrates the efficacy of these requirements. Founded in 1838 as a locksmith shop, the company now makes 110,000 products and has 1,200 employees in its 36 European

branches. Schachermayer is committed to the city's healthy growth and was the first company in Linz to build a green roof. Its complex now boasts almost 115,000 square feet of green roof, nine percent of the city's estimated 1.3 million square feet of green roof coverage. Like the owners of Casa Bauträger *(Building Case Study 2),* Schachermayer financed the installation costs without municipal financial assistance.

COURTESY OF MAGISTRATE OF LINZ

27 | Augustenborg Botanical Garden

LOCATION
Malmö, Sweden

DATE
2001

CLIENT
Malmö City

ARCHITECTS
Gisli Kristjansson

Taddele Gebrevol

LANDSCAPE ARCHITECT
Pär Söderblom

ENGINEER
Per Nyström

SYSTEM MANUFACTURER
Veg Tech AB

GREEN ROOF SIZE
95,000 s.f.

SOIL DEPTH
1 in.-5.25 ft.

Augustenborg's 2-acre Botanical Roof Garden is the largest green roof in Scandinavia and the primary green roof research and testing site in Sweden. The project was built in 2001 atop a complex of industrial buildings in Malmö, near the border with Denmark. It was financed by EU Life and by the Swedish Ministry of the Environment. The research, conducted in cooperation with four universities, focuses on plant development, water retention and runoff, planting methods, soil and drainage, gradients, noise reduction, energy consumption, and roof lifespan.

The plots are used for creative design displays as well as research. Each rooftop section hosts a variety of short-term demonstration gardens. In 2003, four architects showcased accessible intensive gardens for community and healing spaces—models that could be replicated in hospitals, homes for the elderly, and community centers. The designs were both functional and aesthetically innovative: a forest of climbing plants, a rolling hillside, and a relaxing herb garden.

Malmö is an industrial city and suffers frequent flooding, making stormwater management a particular focus of the initiative. The green roof research, and the garden itself, is part of an effort to transform Malmö into an "Ecocity" (*Building Case Study 24*). The local university runs an annual course on the value of green roofs in city planning, and the city is taking steps towards sustainability through alternative energy use, waste management, and increased public transportation.

28 | Milwaukee Metropolitan Sewerage District

LOCATION
Milwaukee, Wisconsin

DATE
2003

CLIENT
Milwaukee Metropolitan Sewerage
District

LANDSCAPE ARCHITECT
Stephen McCarthy

GREEN ROOF CONSULTANT
Weston Solutions Inc.

SYSTEM MANUFACTURERS
ABC Roofing Supply Inc.
Weston Solutions Inc.

GREEN ROOF SIZE
3,800 s.f.

SOIL DEPTH
4 in.

After being cited by the federal government for over a hundred violations of the Clean Water Act, including dumping almost one billion gallons of raw sewage into local rivers and Lake Michigan, the city of Milwaukee has been actively working to reduce the frequency of combined and sanitary sewer overflows. As part of a regional program to promote best management practices (BMP) for urban stormwater maintenance, the Milwaukee Metropolitan Sewerage District (MMSD) installed a green roof on forty percent of its 4-story downtown headquarters.

The $69,000 modular roof system consists of 435 containers of native species and was chosen for its easy installation and the flexibility to adjust and rearrange the modules as needed. The containers and connecting walkway are made of recycled plastics. The roof is planted with prairie species found on dry bluffs in southwestern Wisconsin. The bluffs are known as "goat prairies," and offer native plants only a thin layer of soil over bedrock—harsh conditions similar to those found on extensive rooftops. The roof was

planted with second-year plant plugs, later supplemented with grass seed collected from a bluff.

Monitoring devices on the roof measure runoff volume on both its planted and unplanted sections. The goal of MMSD is to demonstrate the feasibility of a green roof as a BMP, test the rooftop viability of various plant species, and raise public awareness of green roof technology.

COURTESY OF BOB KUEHN

29 | Montgomery Park Business Center

LOCATION
Baltimore, Maryland

DATE
2002

CLIENT
Himmelrich Associates Inc.

ARCHITECT
Werner Mueller DMJM

ARCHITECT
Notari Associates

ENGINEER
Morabito Consultants

GREEN ROOF PLANT SUPPLIER
Emory Knoll Farms

GREEN ROOF CONSULTANT
Katrin Scholz-Barth, formerly of Hellmuth Obata + Kassabaum

WATERPROOFING AND ROOFING CONSULTING
David Hawn, Dedicated Roof and Hydro-Solutions

GREEN ROOF SIZE
30,000 s.f.

SOIL DEPTH
2.5-3 in.

The long-abandoned Montgomery Ward catalog warehouse in Baltimore reopened in 2002 as the Montgomery Park Business Center, home to the Maryland Department for the Environment (MDE). The development transformed a nearly one hundred percent impervious brownfield into a model for stormwater management The building's 30,000 square foot extensive green roof is part of an integrated stormwater management system, which also includes a 30,000-gallon underground rainwater cistern, recycled porous pavement, and bioretention areas in the parking lot.

The state-issued Request for Proposals (RFP) specified that prospective site development bids meet green building guidelines similar to the LEED standards, and was the first RFP in Maryland to focus on sustainable design. The warehouse's original 1925 design incorporated sustainable principles such as durable building materials, onsite transportation, natural lighting, and employee leisure facilities. The developer who won the contract approached the redevelopment similarly, keeping in mind both social amenities and the environmental requirements of the RFP.

The construction process for the green roof was problematic at times, due to tight scheduling around a fixed move-in date for the MDE. Several parts of the warehouse were under renovation at once, which proved difficult when falling glass and debris from window replacement damaged the green roof waterproofing membrane below. The plant plugs were kept on a nearby tar roof to await installation, where they endured significant stress from drought, heat, and wind. The waterproofing membrane was repaired, and as testament to the hardiness of green roof plants, most survived—and spread into a thriving cover after the first year.

COURTESY OF KATRIN SCHOLZ-BARTH CONSULTING

30 | Somoval Garbage Treatment Plant

LOCATION
Monthyon, France

DATE
1997

CLIENT
Somoval

ARCHITECT
S'PACE

SYSTEM MANUFACTURER
Soprema

GREEN ROOF SIZE
165,000 s.f.

SOIL DEPTH
2.5-3 in.

The end result of a contentious debate over the location of a garbage treatment plant outside of Paris was some unusual features on the final design. The Somoval treatment plant in Monthyon, less than a mile from the capital, has a 4-acre hilly rooftop designed to help the structure blend into the undulating French countryside.

The dispute over the treatment plant revolved around issues of accountability, and heightened tensions between urban and rural residents and governments. The eventual solution was an architectural competition to create a design that reduced pollution and allowed the building's appearance to merge with the surrounding wheat fields. French architecture firm S'PACE, renowned for developing large scale industrial projects on sensitive sites, won the competition with a design that reconciles building and landscape. The "tundra roof"—as the manufacturer calls it—is a vibrant green roof planted with sedum and moss, whose bright summer red turns to green in the winter. The green roof insulates the plant, while a small landscaped forest nearby helps mitigate emissions. The plant's operations are also ecological, and include experimentation with sustainable methods of waste treatment, such as clean incineration, recycling, and composting. The project is a model for new ways in which industry can clean water, reuse waste, and unite development with nature.

COURTESY OF S'PACE

31 | Valdemingómez Recycling Plant

LOCATION
Madrid, Spain

DATE
1999

CLIENT
Vertresa-RWE Process

CLIENT
Madrid City Council

ARCHITECT
Abalos & Herreros

LANDSCAPE ARCHITECT
Fernando Valero

ENGINEER
Obiol y Moya

ENVIRONMENTAL IMPACT RESEARCH
Javier Ceballos

SYSTEM MANUFACTURER
AIMAD

SYSTEM MANUFACTURER
ZinCo GmbH

GREEN ROOF SIZE
161,458.7 s.f.

SOIL DEPTH
2.8 in.

The Valdemingómez recycling plant outside of Madrid was built not to last. The building was designed to be dismantled and recycled in 25 years, the estimated working lifetime of the plant. The inaccessible red sedum roof will protect the roof membrane for that period, while keeping the offices below cool through hot Spanish summers.

The body of the building is made of polycarbonate, a cheap and recyclable plastic, which, the architects claim, can be taken apart with a screwdriver and wrench. The green roof system was the same price as a traditional roof, but contains no polyurethane or polyethylene—petroleum products usually found in roofing materials. The roof design features small hills that mimic the surrounding landscape. Four skylights at opposing angles light the offices below. A system of canals divides the cover into 11 sections and drains stormwater runoff.

Mushrooms now grow alongside the sedums—an organic growth that matches the architects' "new naturalism" philosophy of simple, inexpensive, low-impact architecture. The Valdemingómez recycling facility is built to be consistent with its function, fully acknowledging that its own materials will one day be recycled too.

COURTESY OF JORDI BERNADÓ

32 | Ford Rouge Center, Truck Plant

LOCATION
Dearborn, Michigan

DATE
2003

CLIENT
Ford Motor Company

ARCHITECT
William McDonough + Partners

ENGINEER
Arcadis Giffels

DESIGN CONSULTANT
McDonough Braungart Chemistry

STORM WATER CONSULTANT
Cahill Associates

CONSTRUCTION MANAGER
Walbridge Aldinger

GREEN ROOF CONSULTANT
Xero Flor America

RESEARCH SUPPORT
Michigan State University

SYSTEM MANUFACTURER
Xero Flor America

GREEN ROOF SIZE
454,000 s.f.

SOIL DEPTH
2 in.

The largest green roof in the world is on top of the Ford Motor Company's new truck assembly plant in Dearborn, Michigan, and is a key component of the revitalization of Ford's historic River Rouge factory site. Over the course of its 90-year history, the Rouge Center was stripped of vegetation and covered with buildings, rail lines, and parking lots. The soil is contaminated and the company has been cited for violations of the Clean Water Act. Since 1999, Ford has invested billions of dollars to reclaim the area and experiment with sustainable construction techniques.

The 10.4-acre extensive green roof covers almost half of the new assembly plant. Its 2-inch depth is projected to retain 447,000 gallons of runoff annually, amounting to fifty percent of the county's annual rainfall, or approximately one gallon per square foot per year. The 600-acre integrated stormwater management system also includes porous paving and 22 acres of wetlands, and is expected to save the company $35 million.

A consultant to the project suggests that the roof will recreate twenty-five percent of the habitat of an undisturbed greenfield—a twenty-five percent improvement over existing conditions. The roof is expected to improve air quality by forty percent by absorbing dust and breaking down hydrocarbons. Current predictions indicate that energy use at the plant will decrease by seven percent due to the insulating effects of the green roof and new trellises. A new visitor's center will provide a view of the green roof and information about the site remediation. The green roof at River Rouge attests to Ford's efforts to explore ways of greening their facilities and products.

COURTESY OF FORD PHOTOGRAPHIC/ WILLIAM MCDONOUGH + PARTNERS

33 | Possmann Cider Company

LOCATION
Frankfurt, Germany

DATE
1993

CLIENT
Possmann Company

LANDSCAPE ARCHITECT
Siegfried Ziepke

ENGINEER
Werner Volkmar Possmann
Dieter Hach

GREEN ROOF SIZE
32,292 s.f.

SOIL DEPTH
N/A

The Possmann Company, a German apple cider maker, uses rainwater collected on its roof to keep its cider cool during fermentation. Cider is usually fermented in large underground holding tanks cooled with running water, which then drains to the sewer, but Germany's high water and runoff removal fees make this approach costly.

In order to modernize its factory and lower production costs, Possmann added a green roof in 1993. The roof uses the same type of "multi-functional wetland plants" as the KPMG roof (*Building Case Study 3*) to create a closed loop cooling system. Rainwater collected in the wetland circulates into the factory to cool the tanks and then back to the roof. The plants thrive in the warmed water, and the shady root zone cools the water again before it flows back to the factory. Water loss from regular green roof transpiration is compensated for with water retained in the parking area, which is pumped to the roof for natural filtration and treatment.

The roof cooling system has saved money in runoff fees, as well as $6,000 in annual cooling costs. The roof has also helped transform Possmann's image, by sparking numerous articles in the national and international press about its cost-effective, environmentally-friendly manufacturing process. The wetland roof is a favorite location for local birds and bird watchers. A 1999 study also showed that over 20 new species of plants had established themselves.

89

34 | John Deere Works

LOCATION
Mannheim, Germany

DATE
2003

CLIENT
John Deere

ARCHITECT
John Deere

LANDSCAPE ARCHITECT
John Deere

ENGINEER
John Deere

Green Roof Consultant
Ulrich Zens

GREEN ROOF SIZE
450 s.f.

SOIL DEPTH
2 in.

John Deere Works in Mannheim, Germany produces a great deal of wastewater in its manufacturing and assembly operations. Until recently, the company sent this wastewater to a treatment plant, which discharged it into the municipal sewer system for a substantial fee. In order to reduce water management costs, the company decided to experiment with using a constructed wetland treatment system to handle wastewater onsite. There are over 5,000 constructed wetlands in Germany, used primarily to treat residential and municipal wastewater in

areas where a connection to the central sewage treatment system is too costly. Wetland plants clean and filter water naturally: microorganisms on their roots and in wetland soil absorb and break down contaminants, metabolizing them into nutrients.

Constructed wetlands typically require a large land area, which was not available at the Mannheim factory. Instead, a system was devised to install the wetland on a flat rooftop. In order to limit weight and avoid structural modifi-

cation to the existing roof, the wetland was designed with no soil base. The plants grow in a hydroponic system that is less than 2 inches deep. The designers have found that a combination of sedges, rushes, and irises are most effective at breaking down carbon and nitrogen compounds, and can accumulate phosphates and heavy metals. Researchers are conducting ongoing experimentation to determine maximum absorption capacity and the impact of heavy metals on plants.

COURTESY OF JOHN DEERE

35 | Zurich Main Station

LOCATION
Zurich, Switzerland

DATE
2002

CLIENT
SBB Swiss Federal Railways

ARCHITECT
Pool Architect

LANDSCAPE ARCHITECT
Ernst Basler + Partner Ltd.

ENGINEER
Dobler Schällibaum + Partners Ltd.

ENGINEER
TBF + Partners Ltd.

SYSTEM MANUFACTURER
Sarna

GREEN ROOF SIZE
107,640 s.f.

SOIL DEPTH
1.6-7.9 in.

Engineers designing the expansion of the main station in Zurich's densely developed downtown faced the daunting challenge of severely limited available land. In 2000, they hit another, seemingly insurmountable, roadblock. The sandy area earmarked for construction of four new rails and passenger and cargo buildings was home to several protected endangered species, including the wall lizard, a rare grasshopper, and several rare species of bees. Swiss law mandates

that the habitat of endangered species on federal land cannot be destroyed unless it is replaced. Determined to proceed with the crucial expansion, environmental planners for the City of Zurich suggested an innovative solution: recreate the lizard and insect habitat on the roofs of the new buildings.

The roof design includes little hills carved out of the dry substrate, creating shelters where the

animals can take refuge from the cold. Plantings include specialized vegetation to attract insects. A network of connected pathways of the same sandy substrate leads from the ground to a planted wall, in order to attract and guide insects to the three low-lying roofs. After only a year, the vegetation has filled in and lizards and insects have colonized the new habitat. The insulation of the green roof has also provided significant cooling for the station during hot summer days.

COURTESY OF ERNST BASLER + PARTNERS LTD.

36 | Laban Dance Centre

LOCATION
London, England

DATE
2002

CLIENT
Laban Dance Center

ARCHITECT
Herzog & de Meuron

LANDSCAPE ARCHITECT
Herzog & de Meuron

ENGINEER
Ove Arup

SYSTEM MANUFACTURER
Trocal

GREEN ROOF SIZE
4,305.6 s.f.

SOIL DEPTH
5.1-6 in.

Sometimes known as "power station birds," the endangered black redstart thrives in the barren former industrial sites known as brownfields. These areas are often sparsely vegetated and rich in insects and seeds. Prior to the construction of the Laban Dance Centre on a brownfield along the Thames, ecologists found the site to be an important redstart habitat. After lengthy and contentious negotiations, environmentalists, developers, and community groups reached a compromise by reproducing the redstart habitat on the dance center's roof.

Similar to the Zurich train station roof (Building Case Study 35), the Laban brown roof is made from debris found onsite. Brick and concrete are stacked unevenly onto the flat roof to replicate the irregular brownfield. The rubble provides good drainage, and native seeds will colonize the roof over time.

The dance center has been a key component in the renewal of the surrounding neighborhood, and the building has won several awards for its elegant glass and plastic polycarbonate architecture. Several similar roofs have now been built along the Thames to reconcile the needs of community development with biodiversity conservation, while revitalizing brownfield areas that might otherwise be abandoned (Municipal Case Study: London).

COURTESY OF GYONGVER KADAS

37 | Rossetti Bau

LOCATION
Basel, Switzerland

DATE
1998

CLIENT
City of Basel

ARCHITECT
Herzog & de Meuron

LANDSCAPE ARCHITECT
Stephan Brenneisen

GREEN ROOF CONSULTANT
Stephan Brenneisen

PLANT PROVIDER
Fisch Gartenbau

SYSTEM MANUFACTURER
Biber Dach AG Basel

GREEN ROOF SIZE
13,500 s.f.

SOIL DEPTH
3-20 in.

While British developers were considering a rubble roof for the Laban Dance Centre *(Building Case Study 36),* researchers in Switzerland were also exploring ways to use green roofs to mitigate habitat loss. The Rossetti Building, the modernist structure that houses the Institute for Hospital Pharmaceuticals in Basel, is built on a brownfield roughly 300 feet from the Rhine River. Although the building did not displace the original ecosystem of the nearby riverbank, University of Basel professor and habitat remediation expert Dr. Stephan Brenneisen

argued that the building's impact would be reduced dramatically by the addition of a green roof. The roof is now a thriving habitat for spiders, beetles, and grasshoppers.

The roof has two levels, with local soils of varying depths. The designers planted minimal seed, intending that native species would colonize the area naturally. The habitat has attracted over 50 varieties of beetle and some unexpected spiders, including the rare eight-eyed jumping

spider, which usually prefers open land and low vegetation, and the wasp spider, which likes high, strong grasses in which to spin its web and catch grasshoppers. The presence of these species, which are rarely found outside of riverbank habitat, has been a surprising result of the research on the Rossetti roof. With its success in attracting diverse species and pragmatic approach to habitat restoration, the habitat roof has become a model for urban ecological development throughout Europe.

COURTESY OF STEPHAN BRENNEISEN

38 | Bird Paradise

LOCATION
Basel, Switzerland

DATE
1999

CLIENT
University Hospital of Basel

LANDSCAPE ARCHITECT
Stephan Brenneisen

GREEN ROOF SIZE
20,000 s.f.

SOIL DEPTH
3.5 in.

Instead of the usual grey city rooftops, patients at University Hospital of Basel enjoy a view that more closely resembles a sandy riverbank inhabited by migratory birds. Every building in the hospital complex has a green roof, but the roof on the Clinic One building was designed as a test site for urban bird observation and habitat creation. The 20,000 square foot roof is covered with sandy loam and gravel from the nearby riverbank, shaped into the small hills and perches that are the preferred terrain of insect-hunting birds. Some native grasses were planted when

the roof was new, but most of the vegetation has grown from seeds deposited by birds.

The Clinic One roof is frequented by some unexpected species, including black redstarts, wagtails, rock doves, and house sparrows, which typically prefer mountains, fields, and rivers to cities. Not only do these rural birds visit the hospital's roofs more frequently than green roofs in the suburbs or near agricultural lands, but common city birds, like tits and blackbirds, are rarely seen there. The explanation for this sur-

prising phenomenon is that bird species use green roofs selectively according to their natural preferences. In dense urban areas where greenery and food are rare, migratory birds select the rooftops, which resemble their home habitats, for feeding and nesting. While the Laban Dance Centre's rubble roof (Building Case Study 36) compensates for habitat destroyed during development, the Clinic One roof demonstrates the possibility of creating new habitats for migratory species and species whose homes have become scarce elsewhere.

COURTESY OF STEPHAN BRENNEISEN

39 | Centro de Información y Comunicación Ambiental de Norte América (CICEANA)

LOCATION
Mexico City, Mexico

DATE
2001

CLIENT
CICEANA

LANDSCAPE ARCHITECT
Raul Gomez Porras, CICEANA

SOIL SPECIALIST
Raul Gomez Porras, CICEANA

PLANT SPECIALIST
Raul Gomez Porras, CICEANA

CONTRACTOR
Dr. Gilberto Navas, University of Chapingo

COORDINATORS
Cristina Garcia Gomez, CICEANA
Hazett Cervantes Morales, CICEANA
Alejandra Juarez Miranda, CICEANA

GREEN ROOF SIZE
5,812.6 s.f.

SOIL DEPTH
7 in.

El Centro de Información y Comunicación Ambiental de Norte América (CICEANA) represents an innovative community response to Mexico City's increasingly dense development. The city expects to have less than 5.5 square feet of green space per capita by 2005, and the Mexican government offers few protections to safeguard native ecosystems from deforestation and agricultural development. Open green space is a commodity usually reserved for the upper echelons of Mexican society.

The nonprofit CICEANA promotes the creation of sustainable green space in Mexico City. Its unique green roof is a living replica of the area's rapidly vanishing native ecosystem, as well as a novel demonstration project for urban agriculture. Through a coalition with local universities, CICEANA created a living laboratory to explore both ecosystem restoration and small-scale food production. The roof houses climate and biodiversity testing equipment alongside a 7500 square foot (750 square meter) agricultural test plot with agroponics, composting, and greenhouses.

The majority of the rooftop is dedicated to reproducing the Pedegral de San Angel, a rapidly declining ecosystem native to the Mexico City region. In an effort to preserve the ecosystem's biodiversity, CICEANA planted over 25 species from the Pedegral in a special area on the roof. A number of other native sedums, cacti, and agaves complete the rooftop landscape. Unique in the scope of its concept and design, CICEANA's project presents a variety of replicable models for green roof use, from biodiversity preservation to food production.

COURTESY OF CICEANA

40 | Orchid Meadow

LOCATION
Zurich, Switzerland

DATE
1914

CLIENT
City of Zurich

GREEN ROOF SIZE
100,000 s.f.

SOIL DEPTH
8 in.

Some of Switzerland's most biodiverse meadowland is on the Wollishofen plateau outside of Zurich—atop the roof of an almost century-old water filtration plant. The Moos filtration plant, built in 1914, was one of the first reinforced concrete buildings in the region. In order to keep the water cool during the filtration process, the building was topped with layers of sand, gravel, and 8 inches of topsoil from the surrounding farmland. The substrate has mixed over the last ninety years to become a hospitable environment for local flora and fauna.

Over 170 species flourish on the 7.4-acre roof today, including nine orchids that are now rare or endangered on the rest of the plateau. Most impressive of these are the 6,000 individual specimens of *Orchis morio*, an orchid species thought to be extinct in the area around Zurich. The Moos rooftop reflects the richness of species found in the region at the beginning of the twentieth century.

The vegetation and substrate has also kept the roof waterproof since its construction. Minor renovations have been limited to the edges of the structure. The regional government is considering designating it a conservation site to protect the unique botanical wealth and history of the roof meadow.

COURTESY OF STEPHAN BRENNEISEN

Municipal Case Studies

"…In its need for variety and acceptance of randomness, a flourishing natural ecosystem is more like a city than like a plantation. Perhaps it will be the city that reawakens our understanding and appreciation of nature, in all its teeming unpredictable complexity."
Jane Jacobs [1]

Imagining the City
Urban Ecological Infrastructure
By Joel Towers

Potsdamer Platz, Berlin

COURTESY OF MANFRED KOHLER

One of the intentions of this book is to reveal different physical types and cultural meanings of vegetative roofs. Among these is green roofs as a bridge between the traditionally distinct practices of landscape architecture and architecture. The intersection of these disciplines is a useful place to begin exploring how designers understand green roofs and how the resulting roofscapes enter the public imagination and impact the physical environment. Understanding green roofs as both land-form and building-form will pave the way for a broader reading of vegetative roofs as part of a larger "ecological infrastructure" operating between urban spatial form, social process, and environmental function.

From an architectural perspective, green roofs may be seen as products of a tectonic tradition which sees modern architecture as a craft of structure and construction as much as a discipline of space and form.[1] Here the specific historical reference is to earth shelters, a form of

dwelling humans have occupied for millennia. While reasons for earth sheltered habitation vary, the thermal insulation they provide and the regionally specific building and cultural traditions they express (especially a respect for and physical connection with the land) certainly account for much of their continued presence. Green roofs are a contemporary translation of those traditions through modern materials and technologies. Having entered the professional architect's vocabulary, they are now designed and specified much in the same manner as traditional roofs, complete with warranties for properly installed systems. In this sense, the decision to build a green roof usually emerges from concerns about function, cost, and aesthetics integrally bound to the autonomous building itself and the need to put a roof over it. Though perhaps narrowly construed, green roofs from an architectural perspective can be understood, in the most basic sense, as roofing that happens to be landscape.

From the position of landscape architecture, green roofs begin as gardens or amenities where plant selection (native, exotic, drought-tolerant) and soil composition are coordinated with design intention and use. Research and experimentation has resulted in various soil compositions and mat systems that can support rooftop vegetative structures and has allowed landscape architects increased creative freedom in green roof design. However, because the vegetative material from which green roofs are constructed is connected to larger scale ecological questions, the network of considerations for landscape architects is easily expanded beyond any one autonomous building. For landscape architects, green roofs can be seen as resituated earthwork—roofscapes as fields of opportunity under which there happen to be buildings.

Spanning the divide between these two perspectives, green roofs themselves become mediating spaces that negotiate the concerns of both architecture and landscape. As such, they are also sites of imagination providing opportunities for designers to challenge both the autonomy of their disciplinary perspective and the scale (individual, collective, infrastructural, regional, etc.) at which rooftop gardens are typically understood.

In their study of Berlin, Kohler and Keeley attribute the popularity of green roofs in Germany, in part, to an "expression of the country's national environmental consciousness" *(Municipal Case Study: Berlin)*. Extending this idea, it becomes possible to understand green roofs as the spatial translation of a particular cultural perspective. If an "environmental consciousness" is broadly embedded in German social structure, then the individuals, communities, corporations and institutions that constitute that culture can be expected to create expressions of their environmentalism in the constructed landscape, each in their own way and at scales consistent with their agency and theater of operation. Individual homeowners may build green roofs for reasons related to their personal property while corporations may see vegetative roofs as part of their corporate responsibility. Those landscapes, in turn, influence how people understand the world around them. Constituted by and constitutive of the "consciousness" that Kohler and Keeley identify, green roofs are both inspiration and physical structure. The idea of a green roof reinforces a perception of ecological awareness while its presence impacts ecological function. Seen in this way, green roofs offer a compelling physical translation through design of what the anthropologist Arjun Appadurai might have meant by "the imagination as social practice."[2]

From this perspective, a number of questions arise as to how green roofs can be creatively considered in an urban context. Are urban green roofs a sign of changing perceptions about the city? Do they indicate an extension of nature within the city or do they point to a commingling of the natural and built environments in a manner that might lead toward increased resilience and a more integrated socio-natural relationship?[3] How do they impact the dynamics of the urban organism?

The scale and position from which these questions are explored affect how green roofs are seen and understood. Urban green roofs as individual entities may be viewed as extensions of private space. They may also be seen as providing specific, localized benefits to the operation of buildings. Collectively, however, they have the capacity to impact urban ecology and, therefore, may also be understood in infrastructural terms. Mikami's study of Tokyo is instructive of this scalar shift, examining how the aggregation of green roofs in that city are seen as part of an emerging infrastructure that has the potential to mitigate urban heat island effect *(Municipal Case Study: Tokyo)*. The question here is not whether one scale or another is the correct scale, but rather that a multiplicity of scales will result in a variety of meanings for green roofs.

As a result, green roofs acquire social capital in different ways (often related to scale) that may or may not result in their widespread application in urban environments. Understanding this calculus is essential if green roofs are to be one tool in the larger project of shifting the post-industrial city towards ecological (natural and social) sustainability. This shift is in its formative stages and constitutes a possible alternative urban future desperately in need of exploration. There is much work to be done and as the geographer David Harvey reminds us: "[T]he integration of the urbanization question into the environmental-ecological question is a *sine qua non* for the twenty-first century. But we have as yet only scraped the surface of how to achieve that integration across the diversity of geographical scales at which different kinds of ecological questions acquire the prominence they do."[4]

In addition to their imaginative potential, green roofs are also physical spaces. This book is a testament to the remarkable variety of form and material composition they attain. It is clear in looking at the wetland roof on top of the John Deere factory in Mannheim, Germany or the wildlife habitat restoration projects in England and Switzerland that individual green roofs serve a range of discrete ecological and social functions. One must be careful to avoid the tendency to simply aggregate these distinct roofscapes

and suggest that they necessarily mean more than the sum of their parts. Strategic, wide scale development of green roofs has met with varying degrees of success, as the following municipal case studies attest. Nonetheless, these municipalities have advanced the idea that green roofs can be understood as part of an emerging field of ecological design and construction, which leads to the questions: Can such a thing as an "urban ecological infrastructure" exist beyond traditional understandings of urban forestry, park systems, waterways, etc.? If so, how would it acquire social capital? Could urban vegetative roof structures be seen as harbingers of an emerging socio-natural urban ecology, through which cities assign value to human interactions with the natural world on an infrastructural scale? This is certainly one way to read Frith and Gedge's assertion that "[t]he most important catalyst of green roof construction in Britain in the past five years has been the drive to reconcile biodiversity conservation with urban renewal" *(Municipal Case Study: London)*.

However, if green roofs are to be evaluated as a component of an ecological infrastructure, they must cover enough of the city to have a measurable impact on microclimate, energy, and material flows. Given sufficient acreage, green roofs of varying sizes, functions, and designs would constitute a mosaic of inter-related green spaces—individual ecological patches whose benefits could be greatly multiplied to the point of producing larger scale transformations of urban ecologies.[5] Operating in this complex infrastructural fashion, green roofs would attain a social relevance that would produce a feedback loop reinforcing their deployment across the urban landscape.

Should this occur, green roofs might, someday, be as common and essential as sidewalks. Which is to say they would become part of the urban lexicon; no longer extraordinary as spatial form but rather a matter of social process. The vast array of projects in this book makes it possible to begin to glimpse a future in which building and landscape are reflexively integrated and in which urban ecologies—complex relationships between social and natural systems—predominate.

01 | Berlin
Green Roof Technology and Policy Development
By Manfred Kohler and Melissa Keeley

COURTESY OF MANFRED KOHLER

The view from Berlin's Reichstag—a building at the center of Germany's turbulent history, former home to the Imperial government, set ablaze during the Nazi regime—is not of a gray expanse of asphalt and chimneys, but of a sea of greenery thriving atop the roofs of surrounding federal buildings. Berlin has seen substantial growth since reunification, which has brought new construction, along with new green spaces, from parks, gardens, and tree-lined streets to an extensive network of green roofs like those visible from the Reichstag. Green roofs have caught on not only in Berlin, but throughout Germany. In 2001 alone, fourteen percent of flat roofs on new buildings in Germany were green roofs, accounting for approximately 145 million square feet (13.5 million square meters).[1]

The popularity of green roofs in Germany, particularly in the heart of Berlin, is just one expression of the country's national environmental consciousness. In the 1970s, research scientists

rediscovered and studied Berlin's nineteenth-century green roofs, while the recently-formed environmental movement campaigned for "green" urban renewal. The convergence of these movements created a demand for environmental subsidies, incentive programs, and regulatory structures at all levels of government, which in turn led to the mainstreaming of green roofs in the country today.

The Origins of Modern Green Roofs
Germany's first green roofs were the result of an unintentional innovation in the late 1880s—a period of rapid industrialization and urbanization during which construction in Berlin was booming. Rows of apartment blocks known as "rental barracks" (*Mietskasernen*) were built quickly to house the city's exploding population of workers. Inexpensive tar was the most common roofing material at the time, but it was highly flammable; a fire ignited by one coal spark could destroy whole city blocks at a time. Seeking a

less combustible but still economical alternative, German roofer H. Koch developed a roofing technique using a special kind of tar covered with a layer of sand and gravel. The new method quickly became popular, and roofs of this design were constructed throughout Berlin. The fire- and weather-proofing mixture of sand and gravel also turned out to be an excellent growing medium. Seeds found their way onto the roofs, and plants eventually covered the surfaces, creating naturally-occurring plant palettes that closely resembled those used in modern green roofs.

Koch's green roofs were rediscovered in the 1960s when scientists began to research the unusual vegetation growing on some of Berlin's prewar roofs. Free University of Berlin researcher Reinhard Bornkamm, often called the father of modern green roofs, began to study the plant ecology of Koch's roofs. He eventually built the first modern large-scale green roof project in

Berlin Grunewald
The green roof of the water pumping station Berlin Grunewald is an early example of vegetation used for insulation. This nineteenth-century roof was greened to keep the drinking water cool, and eventually became home to several species of rare lichens. Berlin Grunewald was one of the few intentionally-greened roofs of the period, and remained intact until the roof was removed for building renovation in 2002.

The Schering Pharmaceutical Company
The Schering Pharmaceutical Company headquarters, planted in the 1970s, was one of the first large roof gardens in West Berlin. The intensive green roof is elaborately landscaped with trees, shrubs, bamboo, and many flowering perennial plants, and is visible from several apartment complexes and high-rise office buildings.

COURTESY OF MANFRED KOHLER

Berlin on a building at the Free University.[2] Bornkamm's model combined extensive and intensive technologies, but was eventually dismantled due to lack of funding for upkeep.[3] Bornkamm's students later continued his research, using infrared aerial photographs of West Berlin to identify Koch's other remaining historic green roofs. They found a surprising 50 still intact as of the 1980s—after more than 70 years and two World Wars.[4] Even more surprising was the discovery that the roofs had remained waterproof through the entire life of the buildings. The waterproofing membrane on a normal modern roof is typically replaced every 10 to 15 years, worn out by sun exposure and heat expansion. The only damage found on the nineteenth-century roofs made with Koch's technique was on the zinc coverings of parapet walls and around chimneys—areas that still present waterproofing challenges today.

These discoveries sparked interest from

Germany's scientific and business community. University researchers tested the suitability of various plant varieties for roof growth and root-repellant materials to protect waterproofing membranes. Companies eager to introduce and gain acceptance for new green roof-related product lines began their own experiments and built demonstration projects around the country. University of Hannover professor Hans-Joachim Liesecke compared and assessed these green roof systems, offering third party product evaluation for the growing industry. In 1975, a group of scientists, contractors, gardeners, and government representatives founded the Society for Landscape Development and Landscape Design (*Forschungsgesellschaft Landschaftsentwicklung Landschaftsbau* or FLL) to develop universal standards for the construction and quality of green roofs. The establishment of the FLL was a major step towards public acceptance and trust of the technology, and encouraged further research and commercial interest.[5]

These scientific developments took place in an atmosphere charged by the burgeoning German environmental movement of the 1970s. Influenced by international environmental initiatives and concern over population growth and resource allocation, the German government initiated its first environmental program in 1971. The first post-war National Environmental Protection legislation was passed in 1976, and by 1983, the Green Party had seated representatives in Parliament. By the mid-1980s, the influence of the Green Party in the legislature and widespread public support for environmentalism was reflected in political agendas and government policies, including changing ideas of urban development.

Some cities, like Stuttgart, suffered from severe urban climate problems, while environmental issues in Berlin centered around urban renewal, urban quality of life, and pollution. Berlin's green roof initiative began as part of a larger movement to 02

Paul Lincke Ufer
Completed in 1984, the Paul Lincke Ufer urban renewal project experimented with a number of environmental techniques on the existing infrastructure of a neighborhood block in Kreuzberg. Green roofs were tested for air and water quality benefits, as well as insulation value.

COURTESY OF MANFRED KOHLER

beautify the city and preserve its historic districts—an effort that had not been undertaken in the city in 50 years. Following the First World War, the city had invested in the creation of public forests and parks, most of which were then destroyed during the Second World War. In 1961, the East German government erected the Berlin Wall, which divided the city for the next 28 years. In the 1950s and 1960s, the West Berlin government focused on creating new housing for the city's population and reinvigorating the post-war economy. Construction was the first priority, with less focus on greening and environment. The municipal and national governments of the East also made little investment in greening East Berlin.

In the 1970s, the West Berlin government launched an urban renewal plan to replace blocks of old tenement buildings with new high-rise residential complexes. The plan, later derided as "clear-cut renewal," because it resulted in the destruction of entire city blocks, stimulated public resistance to protect the city's characteristic low-rise architec-

ture, affordable housing, and sense of community. Though the issues with this type of development weren't exclusively environmental, there was a grassroots demand to improve urban quality of life through neighborhood "greening."

Building on the public support for a green urban renewal, West Berlin environmentalists campaigned in the early 1980s to "bring nature back into the city" (*die Natur in die Stadt zurückzuholen*). The movement focused on improving living conditions in neighborhoods and apartment blocks through greening projects, including a reduction of paved areas, improvements in public courtyards, and the creation of several green roofs throughout the city.

The movement also led to collaboration between activists and city officials on an experimental project called *Paul Lincke Ufer*. Located in the Kreuzberg district of West Berlin, near the Berlin Wall, the project was an effort to test and demonstrate the aesthetic, environmental, and public health ben-

efits of greening an entire city block. The demonstration included retrofitted green roofs, energy-saving technologies, and the first city recycling trials. Researchers monitored air and water quality and energy use. The success of the project demonstrated the potential of preserving and enhancing existing infrastructure.[6]

Public Policy and Incentives
The West Berlin government began a greening grant program during the trials at *Paul Lincke Ufer* that proved to be one of the most effective catalysts of green roof development in the following years. The Courtyard Greening Program reimbursed residents for roughly half the total cost of green roof installation—between $3.40 to $7.00 per square foot ($37 and $75 or 60 to 120 DM per square meter) of extensive green roof. Intended to increase awareness of new greening techniques and provide test sites for the technology, the grant program helped stimulate the green roof market and drive down costs by subsidizing implementation for interested citizens. Between 1983 and 1997

UFA Film Fabrik

The 1920 UFA Film Fabrik in Berlin Tempelhof has been transformed into an environmental test site. The green roofs are part of the integrated water system, and are monitored for their effect on the efficiency of photovoltaic cells.

Adlerhof Institute of Physics

The Adlerhof Institute of Physics at Humboldt University is built on a groundwater protection area. Under Berlin's landscape plan, the Institute was required to install green roofs to mitigate negative effects on the groundwater. The building is not connected to the public sewer, but manages stormwater onsite with a partially-greened roof, facade greening, a courtyard infiltration pond, adiabatic cooling system, and underground storage containers.

COURTESY OF MANFRED KOHLER

COURTESY OF MANFRED KOHLER

(when city budget restrictions ended the program), approximately 684,000 square feet (63,500 square meters) of extensive green roofs were built on renovated buildings in the center city of West Berlin.[7] In some neighborhoods, smaller grants are still available for green roof projects.

At the federal level, the National Nature Protection Law and the National Building Law mandate that the environmental impact of new development be evaluated and mitigated onsite, if at all possible. This regulation is similar to wetland regulations in the United States, but includes all undeveloped land, not just areas of particular ecological importance. Developers must complete an extensive environmental assessment outlining how environmental impacts will be minimized in their building and site design before they are granted a building permit. Green roofs are a popular mitigation measure because they do not require additional land use, as would planting trees or creating meadow habitat. The national mitigation laws only affect construction on undeveloped greenfields, but

federal law additionally establishes a framework that allows local authorities to develop their own "landscape plans" according to the needs of their town or city.

In order to ensure that the city's green amenities were maintained or enhanced as the city grew more dense, West Berlin implemented a landscape plan called the Biotope Area Factor (BAF) in several of its high-density neighborhoods. The plan assigns numeric targets for the amount of greenery that should exist on each property. The process begins with a block-by-block environmental assessment of a neighborhood to examine current and potential green space, evaluating recreational areas, habitat, biodiversity, and other natural services. Planners then set a greening target level that the property owner or developer must meet to be issued a new building permit. The formula is non-prescriptive, allowing builders to choose from a variety of techniques while ensuring that the green amenities are maintained as the city develops. Green roofs are again a favored

method of mitigation since they do not compromise usable space on the property. The BAF now applies to the unified Berlin, and though it is only legally binding in three neighborhoods, it serves as a guideline for development in the city's other high-density areas.

One of the most compelling incentives for green roof installation is savings on stormwater fees, which also began, indirectly, at the federal level though a split-level fee structure. In 1984, a federal court ruling mandated water utility billing transparency. As a result, many German water utilities formulated new pricing plans. Instead of levying all fees based solely on freshwater consumption, some utilities began to charge for stormwater removal based on impermeable surface area—in effect, the amount of stormwater actually removed from a property. This change significantly influenced the way property owners thought about impermeable surfaces. Because green roofs retain stormwater and delay runoff, they are accounted for in calculations of

Potsdamer Platz; "Daimler City"
A bird's eye view of the Daimler City complex at
Potsdamer Platz reveals that most of the rooftops are
covered with extensive green roof systems. The
green roofs are part of the response to a city council
mandate that the development manage ninety-nine
percent of its rainwater runoff onsite.

Potsdamer Platz; "Daimler City"
Intensive gardens also adorn the lower terraces at
Potsdamer Platz. They were installed to fulfill a
requirement for outdoor courtyards and play spaces.

COURTESY OF MELISSA KEELEY

COURTESY OF DANIEL ROEHR

impermeable surface, and lessen the runoff fee
on the property. A green roof with a 4-inch (10-
centimeter) substrate absorbs about 1.4 inches
(35 millimeters) of water, significantly lessening
runoff.[8] Some U.S. cities, like Portland, Oregon,
are adopting similar models to address their
stormwater management problems.

The split stormwater fee system has had a major impact on the city even in the short time since
its adoption, largely due to new construction in
both the west and east of a reunited Berlin. Before the fall of the Berlin Wall, East Berlin invested little money in urban greening. Conse-
quently, after reunification, former East Berlin
neighborhoods became the focus of intensive
greening efforts. Greening subsidy funds for
Courtyard Greening Programs were concen-
trated for some time in the east.

Most notable are the greening efforts in the re-
construction of Potsdamer Platz. Once a vibrant

commercial center, it was leveled during World
War II and bisected by the Berlin Wall, remaining
a derelict, undeveloped no-man's-land for almost
three decades. After reunification, the city council proposed an ambitious plan to rebuild
Potsdamer Platz as the heart of the unified Berlin—making it for a time the largest construction
site in Europe. The site is anchored by the 17-
building Daimler Chrysler complex, which had to
be built in accordance with the city council's mandate to manage ninety-nine percent of stormwater
onsite. The complex features an artificial lake and
grey water systems, and every building has a
green roof, totaling approximately 430,000 square
feet (40,000 square meters).

Greening subsidies are credited with helping to
spark the green roof market in Germany, and even
today, with the city in a fiscal crisis, some subsidies continue in areas of special need around
the east. For example, the 100 Courtyards program in Berlin-Prenzlauerberg is ongoing, as part

of the city's global Local Agenda 21 Program.[9]
However, due to budget shortfalls at the national,
state, and local levels, many subsidies have been
eliminated. Green roofs are now increasingly promoted through governmental mitigation regulations, fee structures, and permit requirements,
like those used in Potsdamer Platz.

As of 1983, 24 German cities—including Berlin—
had begun subsidy programs to support urban
greening projects.[10] By 1997, a survey of two hundred German cities showed that the number of
cities offering subsidies and other incentives had
almost quadrupled.[11] Half of Berlin's new government buildings have green roofs. The
mainstreaming of green roofs in Berlin and across
Germany occurred through technological innovation and standardization, citizen activism, and
legal support at all levels of government. These
factors have been crucial not only to development of green roofs, but to greener cities and a
more sustainable future for Germany.

02 | Tokyo
Cooling Rooftop Gardens
By Takehiko Mikami

Atago Building
A small square of green amidst a sea of steel, the roof of the Atago Building demonstrates the resourcefulness of Tokyo's building community at incorporating green roofs into new skyscraper design.

COURTESY OF MORI BUILDING COMPANY

After almost six decades of nearly unmitigated growth, only fourteen percent of Tokyo's land area remains green. The city has the lowest green-space-to-impermeable-surface ratio of any major metropolis; the urban heat island effect is causing Tokyo's temperatures to increase at a rate five times faster than global warming. Faced with high density, increasingly intolerable temperatures, and other environmental and public health problems associated with dense urbanization, the municipal government has turned to green roofs as a solution. New green roof laws and regulations have found widespread support in the private sector as developers use elaborate rooftop gardens to distinguish their buildings and attract tenants.

Historic Green Roofs
In the early twentieth century, prominent urban department stores began installing elaborate roof gardens as a way to showcase their status and offer amenity space to customers.

The Mitsukoshi department store built one of the first roof gardens in Tokyo in 1914. Like most of the department store gardens to follow, the Mitsukoshi roof was built with a traditional Japanese garden aesthetic and included plum and cherry blossoms in the spring and colorful maple leaves in the fall. The original garden was damaged during the Kanto earthquake in 1923, but was later rebuilt. Today, department stores commonly install roof gardens, which often feature paths, benches, and playgrounds.

As the city grew, private roof and terrace gardens became popular, and organizations occasionally used roof space for plantings. The oldest extant roof garden in Tokyo is the Asakura Sculpture Museum, constructed in 1936. The building was designed by Fumio Asakura, the founder of Japanese modern sculpture, and included an olive tree planted among the flowers, which is still alive today.

Urban Heat Island Effect
As Tokyo, a temperate seaport city, has become more tropical in the last 25 years due to the urban heat island effect, interest in modern green roofs has grown. The urban heat island was first recorded in Tokyo in 1939, and increased temperatures have become a serious problem since the end of World War II. The city's priority in the aftermath of the war was to rebuild infrastructure and the economy as quickly as possible. Zoning laws and building codes were written with the sole objective of encouraging rapid, dense construction, and as a result, growth went unchecked for decades.

A rising urban population in the 1980s and 1990s fueled a second construction boom. At the same time, Tokyo saw a dramatic increase in fossil fuel usage. As in most contemporary cities, use of fossil fuels increased exponentially after the war, as the city established itself as a global capital for technology and finance. Air condition-

Tokyo Metropolitan Government Assembly Hall
The green roof on the headquarters of the Tokyo Metropolitan Government Assembly is the first official demonstration project sponsored by the city. The quarter-acre roof garden cost approximately $500,000 and features a range of plots, from lightweight sedums to intensive shrubbery landscaping.

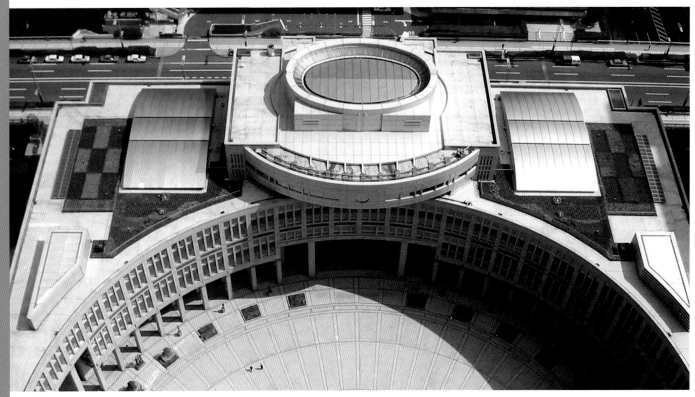

COURTESY OF TOKYO METROPOLITAN GOVERNMENT

ing in central Tokyo office buildings creates enormous energy demand, which must be addressed in order to effectively lower city temperatures—a challenging catch-22.

In the last hundred years, the world's average mean temperature has increased by 1°F, while Tokyo's has increased by 5.2°F[1]—nearly twice the increase experienced in other major cities.[2] According to a Japanese Ministry of Environment report, the number of hours exceeding 86°F on a Tokyo summer day has tripled in the last 20 years. In 2002, there were 29 nights when the temperature did not drop below 77°F overnight—a nighttime temperature which is defined as a "tropical night."[3] Tokyo experienced 18 smog alerts in 2000, one-third more than in previous years.[4] Energy consumption for cooling purposes increased fifteen percent from 1990 to 1998, in turn causing carbon dioxide emissions to rise by six percent. Winters are shorter, and the annual cherry blossom festival has been

rescheduled because spring arrives earlier. The city's biodiversity has changed as well—tropical species such as palm trees and wild parakeets have recently begun to appear. A small outburst of dengue fever followed one unusually hot and humid period—mosquitoes carrying the virus are generally only found in tropical climates.[5] Heat-related illness and death are also on the rise.[6]

Public Policy Solutions

The severity of Tokyo's climate problems—particularly those affecting human health—has forced the municipal government to respond. A 2001 Environment Ministry study found that the high percentage of impermeable heat-absorbing surfaces in Tokyo contributes directly to the city's warming, causing the urban area to receive thirty-seven percent more heat from the ground than the natural environment does.[7] The ministry estimated that the city's high-density building patterns make it impossible to change land use policies at this late date. The study in-

stead recommended widespread greening efforts throughout the city, including tree planting, park expansion, and a proposed 11.5 square miles (30 square kilometers) of green roofs.[8]

Then Ward Greenery Promotion Director Kazuyoshi Kojima implemented a successful trial greening plan in the Shibuya ward, one of Tokyo's densest neighborhoods. He surveyed companies in the central city to assess reactions to a greening regulation. Twenty-nine firms were willing to convert to green roof at their own expense. Encouraged by this success, he developed an informal incentive program that provided a technology discount and free consulting. A subsidy program he began the following year greened 70,000 square feet (7,000 square meters) of rooftop in one year, one-fifth of which were retrofits.[9] Kojima has become a central figure in the private market development for green roofs and has since opened a green roof general contracting company.

Mitsui Sumitomo Insurance Surugadai Building
The Mitsui Insurance building provides a park-like
recreation area on top of its mulit-story structure.

COURTESY OF MITSUI SUMITOMO INSURANCE CO. LTD

In 2001, the Tokyo Metropolitan Government began to amend its Natural Conservation Ordinance. Based on the environment ministry's recommendations and the success of Kojima's work in Shibuya, the new amendments include a mandate for green roofs on new-construction buildings. Private buildings larger than 10,800 square feet (1,000 square meters) and public buildings larger than 2,700 square feet (250 square meters) must green twenty percent of the rooftop or face an annual penalty of $2,000 (200,000 yen).[10] In the first year after the passage of the law, the total net area of green roofs almost doubled—from 564,000 square feet (52,428 square meters) in 2000 to over 1.1 million (104,412 square meters) by the end of 2001. In addition, the Green Tokyo Plan establishes a target of nearly 3,000 acres (1,200 hectares) of rooftop greenery, as part of the effort to increase green space in the city by ten to thirty-five percent.[11] To promote the legislation, the Metropolitan Government also con-

structed a green roof demonstration project on the Tokyo Council Building and other facilities.

The Japanese government is now following Tokyo's lead on green roof legislation. In 2003, the Ministry of Land, Infrastructure, and Transport announced revisions to the national nature conservation regulations, mandating that all new-construction multiple-dwelling houses and office buildings in urban areas green at least twenty percent of their rooftops. The law goes into effect in 2005.

Amenity Roofs

No penalties have yet been issued under the municipal government's green roof mandate, an indication that all buildings have thus far complied with the law. Compliance is part of a strong sense of social and civic responsibility in Japan, but businesses have also found that roof gardens make buildings distinctive, just as they did for department stores in the early 1900s. Many

developers are augmenting simple sedum mat green roofs with amenities to create accessible roof gardens with a variety of plantings and features.

One of the first projects of this type is the ongoing Roppongi Hill development, initiated in 1986 through a government renewal plan for Roppongi Street, one of Tokyo's busiest. Roppongi Hills embodies a new kind of development ethic and aesthetic. Developer Mori Building Company's plan to transform the area into a new urban district included the creation of a "culture city" and a perpendicular "garden city," both open to the public. Designed to bring together Tokyo's artistic and entertainment communities, the culture city is the site of the innovative Mori Art Center, which showcases contemporary art, architecture, design and fashion. The district also includes the Asahi Television tower and the largest Virgin Megaplex theater in Japan. The garden city

Roppongi Hills
The rice paddy on a Roppongi Hills roof epitomizes the creative spirit of Tokyo's green roof development. The plot is one of many gardens in the Roppongi commercial and residential complex.

COURTESY OF MORI BUILDING COMPANY

features a variety of interlocking rooftop pathways leading to different levels of green spaces. The roofs of almost all the skyscrapers in the complex are green, and have features such as stone walls, topiary, a vegetable garden, a rice paddy, and simulated woodland for wildlife.

Some of Tokyo's other accessible green roofs include St. Luke's International Hospital in Tsukiji,

where pathways wind through ornamental gardens, offering quiet retreat for patients, staff and guests; and the roof of the Infrastructure Ministry headquarters in Nagatacho, which spans 18,300 square feet (1,700 square meters) and boasts olive and tangerine trees and expanses of lawn.

The green roof movement in Tokyo has been fueled by rising temperatures in the city and the

environmental and health concerns they create. Government regulation has guided the movement, but its success owes a great deal to the private developers who have invested in intensive gardens. This commitment to civic responsibility and quality of life on the part of city residents will continue to be important to the ongoing greening of Tokyo.

03 | London
The Wild Roof Renaissance
By Mathew Frith and Dusty Gedge

Deptford Creek
Deptford Creek, a former industrial area by the Thames River, is now targeted for the creation of over 40,000 square feet of rubble roofs—as a mitigation measure to protect the habitat of the black redstart. Britain's first rubble roof, on the Laban Dance Center, was built from crushed concrete from the construction site. The roof is not planted, but plants are expected to colonize naturally from deposited seeds. The Creekside Education Center is the site of a second rubble roof.

COURTESY OF DUSTY GEDGE

COURTESY OF DEREK BROWN

The growing green roof movement in London was initially inspired by a bird. The black redstart, one of Britain's most endangered avian species, touched off a battle between ecologists and developers when a brownfield in Deptford chosen for re-development in 1997 was found to be an important nesting site for the birds. Although that site was developed, some measures were taken to replace the habitat. By 2000, designers conceived an innovative "rubble roof" model. The first rubble roof was built on top of the new Laban Dance Centre, just upstream from the earlier site, in 2002. While technically a green roof, its surface is actually more brown than green, to mimic the appearance of the surrounding barren landscape in an artificial habitat for the birds. The Laban green roof was the first of its kind in Britain, where a conservative approach to architecture and planning and a lack of public awareness about their benefits has kept green roofs out of the mainstream. Since the construction of the Laban roof, however, there

has been increased interest in green roofs as advocates tout the biodiversity advantages and push for the greening of urban renewal projects. The Mayor of London has included language on green building technology in recent urban planning documents. With this newfound legislative support, green roofs, and other green building techniques, may help transform London's building landscape.

Deptford Creek
Biodiversity has slowly become a more mainstream urban policy issue in the United Kingdom since the British government became a signatory on the Convention on Biological Diversity at the Rio de Janeiro Earth Summit in 1992. Following the summit, the government initiated the national Biodiversity Action Plan (UK BAP), a wide-ranging strategy for habitat and species conservation. The UK BAP provides a framework for local environmental initiatives through the establishment of local BAPs.[1] In the early

1990s, three reports on the benefits of green roofs for biodiversity conservation were published, though they were not widely disseminated or acted upon.[2]

Since 2000, the British government has begun implementing "urban renaissance"[3] strategies to revitalize towns and cities. With seven million residents, London is the most populous city in the European Union, and one of the most densely built. By 2016, London's population is projected to increase by another ten percent.[4] As part of an effort to manage this growth through astute urban planning, the national government decided to pilot its urban renaissance program in London. The plan's guidelines call for high-density or infill development, with sixty percent of new construction designated for brownfield sites.[5] London has the largest land area of any European city; the plan called for reclamation of former industrial sites for new development as a way to slow the rate of sprawl.

Canary Wharf
London's new financial district, Canary Wharf, features the city's highest concentration of green roofs. The area has become a site for biodiversity testing, as researchers compare the ecology of the sedum mats to that of the rubble roofs on the nearby brownfields.

COURTESY OF MATTHEW FRITH

The biodiversity conservation and urban renaissance policies have come into direct conflict over brownfield development. Despite a poor public image, many brownfields shelter diverse species, including some who are legally protected or endangered.[6] There are even a few species—birds such as linnets and the rare "humble bumble" bee—for which urban brownfields have become primary habitat.[7] Prior to the construction of a new housing development at the former Deptford Power Station, a team of ecologists evaluated the Deptford Creek brownfield and discovered that the area was the habitat of the black redstart, one of Britain's most endangered birds. During the breeding season, the bird is protected under the Wildlife & Countryside Act. Construction and/or demolition can be stopped if it is found to disturb the bird, and developers must mitigate any loss or damage of habitat.[8] As the law only protects the habitat while the birds are nesting, the developers were legally allowed

to begin construction after the season. Despite a lengthy battle to prevent destruction of the redstart's habitat, developers of the Deptford Power Station decided to proceed with the project, refusing to work with ecologists or consider a compromise plan. However, the developers of Greenwich Reach 2000, the subsequent project at Deptford Creek, did agree to be more environmentally sensitive, in order to avoid the delays and protests experienced by the power station developers. They designed the first rubble roof, though it has yet to be built.

Developers of the nearby Laban Dance Centre also chose to work with ecologists, and were the first to build a rubble roof. The Dance Centre roof is a sparsely vegetated, rock-strewn green roof designed to mimic the brownfield below. The roof has helped to raise awareness about brownfield ecology and the potential value of green roofs. Deptford Creek has become a model for reconciling—at least in part—

local ecology with development. It has been featured in the England Biodiversity Strategy[9] and the British government's 2002 sustainable development report.[10] The London Biodiversity Partnership, established in 1996, lists the black redstart as a priority species and the Deptford Creek development as a model for brownfield species conservation. Several months after the completion of Laban, another rubble roof was installed on the Creekside Centre less than 656 feet (200 meters) away. Today at least one million square feet (100,000 square meters) of green roofs are planned for black redstart habitat in London, and the potential of expanding this model for other species is under investigation.[11]

The experience at Deptford Creek has been a political success for the biodiversity movement. Nevertheless, these rooftops may not be sufficient to protect the black redstart from the rapid re-development of brownfield land. The real im-

COURTESY OF MATTHEW FRITH

COURTESY OF G. KADAS

pact of Deptford Creek has been to educate developers about the value of brownfield ecology, and to put biodiversity onto the green roof agenda.

The new green roof movement in London is focused on biodiversity benefits, which differs from the sporadic British interest in green roofs over the last century. Britain's earliest green roofs served practical military purposes—airfield hangars from the mid-1930s were covered with turf camouflage. A few of these still survive, but they were not intended for long-term use. The first deliberate green roof in Britain in an urban context was probably the roof garden of London department store Derry & Toms, built in 1938. In contemporary Britain, the first green roofs served a primarily aesthetic function, as an option for innovative architects, who integrated roof plantings into design.[12] The new green roof movement in London, and throughout Britain, arises instead from growing ecological concerns

brought to light by the plight of the black redstart and by the issues increasingly raised by climate changes.

Public Policy

Prior to the Deptford Creek development, and even since its success, governmental policy in London (and Britain) has largely overlooked green roofs. For example, green roofs are not currently included in the existing standards for sustainable urban drainage systems. There is also a scarcity of available technical guidance—the English translation of the German FLL specifications is not readily available in Britain[13] and the first non-retail guide to green roof construction, *Building Green*, was published quietly by the Nature Conservancy Council in 1993. More recently, the British Council for Offices has developed an advice note on green roofs aimed at the planning and development sector. This and the 1993 version of *Building Green* are now available on the web.[14]

However, London's municipal government is currently working to incorporate green roof policies into new legislation. The Greater London Authority (GLA), a new regional governing body with strategic planning authority over economic development, transport, biodiversity, and energy, presides over urban renaissance development. London Mayor Ken Livingston, who also heads the GLA, supports green roofs as a solution for issues like biodiversity, energy, open space, and aesthetics, and has written policies accordingly. Under his guidance, the Biodiversity and Energy Strategies include limited provisions for green roofs.[15] The GLA refers to green roofs in the London Plan, the regional planning guide that directs implementation of new development. The Plan provides that "wherever appropriate, new development should include new or enhanced habitat, or design (e.g. green roofs) and landscaping that promotes biodiversity, and provision for their management.[16] The Plan will also be supported

Jubilee Park
Situated above the Canary Wharf Underground Station, the Jubilee Park provides a green respite from the surrounding office towers. The curving architecture contrasts with the straight lines of the financial district.

Wood Wharf
Wood Wharf is a large-scale development planned next to Canary Wharf. The master plan provides for green roofs specifically directed towards black redstart conservation.

COURTESY OF G. KADAS

COURTESY OF BRITISH WATERWAYS

by Supplementary Planning Guidance on sustainable design and construction, with specific elements to encourage green roofs.

Green Roofs Around London

Some private development projects have also advanced the green roof movement in London, motivated by both aesthetic and ecological concerns. The developer of London's new financial district, Canary Wharf, simply felt that the city was too grey and designed green roofs as an alternative. He first worked with local green roof companies and ecologists to retrofit green roofs on some of the area's skyscrapers and has since drawn up a green roof action plan for all future and existing buildings in the district. Most of the roofs are traditional sedum mats, but some are more playfully designed. Jubilee Park, for example, is both a park and the roof to the Canary Wharf Underground station, planted with grassy hills around the domed

subway entrances. 54,000 square feet (5,000 square meters) of green roofs were installed in Canary Wharf in 2001—one is slated to be the highest in Europe, atop a 35-story building—and many new buildings in the area will be constructed with rubble or sedum mat green roofs. Though they were initially installed for aesthetic reasons, the Canary Wharf roofs are proving useful to ecologists. Researchers are examining insects on the rooftops to learn more about replicating habitat for brownfield species, while environmentalists use Canary Wharf as an example of success when touting the benefits of green roofs to other developers and contractors.

Another unique development is BedZED—Beddington Zero-Emission Development—an experiment in social housing design that meets the sustainable development objectives of the British government's urban agenda. The build-

ings are made from locally-sourced wood, glass, and concrete, feature curved "sky gardens" on the second floor apartments, and are topped with colored wind-cowls and an extensive green roof. The sedum mats reduce surface water runoff and provide green space for residents.

The most important catalyst of green roof construction in Britain in the past five years has been the drive to reconcile biodiversity conservation with urban renewal. Using existing species protection legislation, creativity, and enthusiasm, proponents have helped to build the first green roofs in London. It is likely that more will be constructed in the next few years in response to the Mayor's interest in the technology. Advocates hope that legislative support and new public and private green roof projects around the city will further encourage a popular ecological movement for green roofs.

04 | Portland
A New Kind of Stormwater Management

By Tom Liptan

Hamilton Apartments
Hamilton Apartments, a Portland Housing Authority building, was one of the first municipal ecoroof demonstration and testing sites. It is planted with low-growing sedums, native wildflowers, and grasses.

COURTESY OF THE CITY OF PORTLAND

Portland is Oregon's largest city and the first in the U.S. to pass legislation promoting the use of green roofs. A 1994 state order to comply with the Clean Water Act compelled Portland's municipal government to reduce stormwater runoff—no easy task in a city where impermeable surfaces cover one-third of the land area.[1] Portland, like many U.S. cities, has a combined sewer system in which rainwater runoff is processed in the same treatment facilities as household waste. The system is prone to overflow during Oregon's frequent heavy rains. The estimated cost of structural alterations to reduce combined sewer overflows (CSO) is more than $1 billion,[2] and runoff reduction is still an important part of the solution. The CSO control plan integrates structural modifications (partial sewer system separation, large tunnels, and wide pipes) with increased surface water absorption strategies. Improved surface water management is achieved through downspout disconnection, porous pavement, planters and swales, infiltration and water gardens, vertical landscaping, trees, and extensive green roofs, called "ecoroofs" in Portland.

Prompted by the success of the first municipally-funded ecoroofs, the city government has implemented policies to encourage their widespread installation.

History of Green Roof Development

Water pollution has been Portland's primary environmental problem since the late 1980s. Most surrounding streams carry the maximum daily load of state-listed pollutants, and six miles of the local Willamette River bottom are listed as a Superfund site. The region's salmon are marked as an endangered species. In addition, there are concerns about groundwater quality and the city's aging sewer system infrastructure. At the core of the problem is stormwater runoff.

Portland began to research ecoroofs as a stormwater management tool in the mid-1990s. In early 1994, Portland Bureau of Environmental Services employee Tom Liptan noticed the label on a bottle of Ecover dishwashing soap in his kitchen. The label read, "made in an ecological

factory, with a grass roof." He was intrigued, and prompted Environmental Services to research grass roofs. Ecover sent a video, and over the next two years Environmental Services staff worked with a number of German and Canadian green roof experts. In 1996, Liptan constructed an ecoroof on his garage to do some stormwater management testing. His retention and flow tests over the next two years confirmed the roof's efficacy. Liptan's data showed that his ecoroof retained an average of thirty-five percent of total rainfall and almost one hundred percent of rainfall from warm season storms. The garage also remained more than 15°F cooler inside than outside at the hottest part of a summer day.

Encouraged by Liptan's results, the city funded an ecoroof demonstration exhibit at the Portland Home and Garden Show. City officials at the show surveyed public opinion on ecoroofs, and found that seventy-five percent of responses were favorable.[3] Portland made its first serious financial commitment in 1999, when two developers ex-

Garage Roof
Tom Liptan built Portland's first ecoroof on his garage to test its efficiency as a stormwater management tool.

Jean Vollum Natural Capital Center
The Jean Vollum Natural Capital Center was renovated with green building technologies, including a 5,123 square foot ecoroof planted with native species. Ecotrust sponsored the renovation as part of its mission to promote a world where people and wild salmon thrive. The 1895 former warehouse was the first in Portland to be LEED gold certified.

COURTESY OF TOM LIPTAN

COURTESY OF THE ECOTRUST CENTER AND SAMUEL BEEBE

pressed interest in constructing ecoroofs on their new buildings. One of these, the 10-story Hamilton Apartments, was the first affordable housing project in the United States to have a green roof. The second, Buckman Terrace, is a market-rate five-story residential building with a 1,500-square foot ecoroof, and the structural capacity to add another 25,000-square foot green roof in the future.

These two buildings were the first municipally-funded green roof test sites in the U.S., and were built with Environmental Services funds from stormwater fees. Data from the first year of the Hamilton and Buckman ecoroofs indicate that water retention is most effective during the short, intense rainstorms common in the summer. The water retention rate in the wet season (October to May) is ten to thirty-five percent, while the dry season summer rate (June to September) is sixty-five to one hundred percent. The study's overall conclusions showed that a 5-inch-deep ecoroof controls stormwater runoff very effectively, absorbing an average sixty-nine percent of annual rainfall.[3,5]

The Hamilton and Buckman research has encouraged Environmental Services to promote ecoroofs

as a stormwater management tool. The government has used a variety of policy structures to support green roof construction. Environmental Services plays an active role in providing technical assistance to developers and architects. This guidance has facilitated green roof development by streamlining the permitting and construction process. A city-funded ecoroof demonstration grant program has funded 12 projects since 1999 and provided information on vegetation and design options. Technical data from the grant projects is documented and is available to anyone interested in ecoroof technology. As part of a public outreach effort, city staff also makes presentations to developers, consultants, and at conferences. The city plans to continue researching ecoroof design, performance, policy, and economic issues.

Public Policy

Most public ecoroof projects were financed by stormwater fees collected by Portland's stormwater utility, which has been in place since 1977.[4] Unlike most U.S. cities, which bill only for water consumption, Portland has a split fee system, which charges separately for water consumption, sanitary discharge and treatment, and

stormwater management. In 2001, the city reformed the stormwater fee into two parts—thirty-five percent for drainage on the property and sixty-five percent for drainage onto the public streets, calculated according to the amount of impermeable surface area on the property. This approach makes landowners responsible for the stormwater runoff volume created by each building on the property. Moreover, a new policy implemented in 1999 requires that developers who create or change more than 500 square feet of impermeable surface must manage stormwater onsite. If conditions onsite do not allow for full water quality treatment and control of the flow rate, developers can either build an offsite facility, or pay an "in-lieu" fee to the city for municipal stormwater management. A green roof reduces the amount of impermeable surface on a building site, thereby reducing stormwater runoff. Portland is working on a thirty-five percent reduction of the "in-lieu" fee for buildings with ecoroofs.

Ecoroofs were recognized as a Best Management Practice (BMP) in the city's stormwater manual in 1999. Portland will soon implement a

Multanomah Building
The Multanomah Building is another city demonstration project, completed in July 2003. Planted with wildflowers, grasses, and sedums, the site is publicly accessible and provides a beautiful view of the city.

One Waterfront Place
One Waterfront Place is built on a former brownfield site and took advantage of the municipal floor-area-ratio bonus. The 12-story building was able to add an additional 50,000 square feet of space because it also includes an ecoroof. South side apartments have access to the ecoroof gardens, while those on the north side benefit from the green view. Bioswales, trellises, and ecoroofs on the office building reduce the impact of the complex on the stormwater system.

COURTESY OF THE CITY OF PORTLAND

COURTESY OF BOORA ARCHITECTS

Clean River Incentive Program that provides a discount on the stormwater runoff fee for properties using certain BMPs. Installation of vegetative techniques, including green roofs, offers one of the highest discounts.[5]

In the early 1980s, Portland had recognized roof gardens as an asset to the urban environment, and had created the "Floor Area Ratio" (FAR) zoning code bonus for buildings that included them. The bonus applies to the dense downtown district, a priority stormwater management area where zoning limits a building's height-to-floor-area ratio. This ratio can be increased if the design includes features that benefit the community, such as art, locker rooms, water features, day care, or ecoroofs. By greening all or part of a roof, a developer can add as many as three additional square feet to the building for every square foot of ecoroof. For ecoroofs that cover up to thirty percent of the roof area, one square foot of bonus is allowed for each square foot of ecoroof; for ecoroof coverage up to sixty percent of the roof area, two square feet of bonus is allowed for each square foot of ecoroofs; and for coverage greater than sixty percent, each square foot of ecoroof allows three square feet of

bonus.[6] The original roof garden bonus was not widely used until the city added ecoroofs (extensive green roofs) to the program in 2001, but in the three years since ecoroofs were included, five projects have taken advantage of the bonus.

The Office of Sustainable Development (OSD), established in 2001 with funds from the city garbage franchise fee, offers further municipal support for green roofs. The office develops and advocates strategies that ensure the availability of natural resources for future generations and promotes fairness and quality of life. OSD has pressed the city to require new public buildings to meet Leadership in Energy and Environmental Design (LEED) standards. LEED criteria ensure that projects engage in water and energy conservation, improve indoor and outdoor air quality, reduce urban heat island impacts, and implement other green design and construction techniques. Many private developers have begun to apply the standards in order to construct better buildings and gain competitive advantage; tenants and homebuyers are willing to pay more for the amenities in a LEED-certified building, including green roofs.

While developers use LEED certification to increase property values, the citizen group Ecoroofs Everywhere is committed to the equitable distribution of green roofs across all income levels. The group creates affordable demonstration projects, secures grants for small-scale developments, and negotiates lower prices with vendors.[7] Ecoroofs Everywhere has helped to democratize green roofs, a valuable enhancement to government green roof incentives.

By January 2004, over 30 green roofs were built or under construction in Portland, including five on government buildings. The City of Portland has raised awareness of the benefits of ecoroofs so effectively that the private sector has begun to construct them on its own. The development community is beginning to see economic opportunities in the ecoroof movement. Many private citizens are installing ecoroofs on garages, sheds, and houses. Through the work of government agencies and nonprofit groups, ecoroofs are becoming a mainstream technology. Portland's government is committed to providing technical assistance and incentives to green the city and to create a sustainable, livable community for the future.

05 | Chicago
Towards a New Standard of Green Building
By City of Chicago Department of Environment
and City of Chicago Department of Planning
and Development

Chicago City Hall
The Chicago City Hall is planted with over 150 species of plants and is a test site for temperature monitoring and biodiversity research. Readings on a 90°F afternoon in August 2001 averaged 100°F on the planted sections of the City Hall roof and 170°F on the surface of the black tar roof of the adjacent County Building.

COURTESY OF THE CITY OF CHICAGO AND CONSERVATION DESIGN FORUM

Mayor Richard M. Daley came to office in 1989 committed to making Chicago "the greenest city" in the country. Realizing his vision has required the collaboration of every level and department of city government on a wide variety of green strategies, with green roofs at the forefront of the plan. After seeing the benefits of the extensive green roof network in Hamburg, Germany, Daley incorporated green roofs into his greening plan. His office called for investigations into green roof benefits and potential policy incentives, as well as demonstration projects, including the first municipal green roof atop Chicago City Hall. The mayor's enthusiasm for sustainable development as a solution to the severe urban environmental problems facing the city has catalyzed a boom in green roof development. As of 2004, Chicago claims more green roofs than any municipality in the United States.

Chicago's Heat Island
In 1997, Chicago was selected to participate in

the Environmental Protection Agency's Urban Heat Island Initiative, a project to research and reduce the urban heat island effect in cities throughout the U.S. The study found that ozone levels in Chicago were between 180 and 280 parts per billion, well above the EPA's safe upper limit for ground-level ozone of 80 ppb. A Northwestern University study conducted at the same time found that the city's high ozone levels also affected the suburbs.[1] Throughout the Chicago region, air temperatures were found to be 3°F to 5°F higher than in the outlying countryside.

The heat wave during the summer of 1998, Chicago's hottest summer on record, underscored the Urban Heat Island Initiative's findings. The city experienced a series of major power outages and a corresponding rise in heat-related illness and death. The Department of Environment began working with the EPA to install light-colored roofs and replace asphalt with

green space, as recommended by the 1992 EPA report *Cooling Our Communities: A Guidebook on Tree Planting and Light-Colored Surfaces*. The following year, the city's heat-reduction measures took a different shape when Mayor Daley visited Chicago's sister city of Hamburg. He was impressed with Hamburg's many green roofs and how effectively they lowered city temperatures. Soon after the mayor's trip, a lawsuit settlement resulting from the 1998 power outages forced Commonwealth Edison, the electric utility provider, to pay the city $100 million over four years. Mayor Daley directed a portion of the funds towards green roof development and heat management research grants. To demonstrate the city's commitment to innovative greening strategies, the mayor further proposed that City Hall have a green roof.

The roof of City Hall was completed in early 2001. It includes over 150 species of trees, vines, grasses, and shrubs, many native to the region.

900 North Kingsbury
This former Montgomery Ward catalog building was converted into commercial and condominium space. The developer removed the center portions of the building's roof in order to create a courtyard space, which was retrofitted with a green roof as an amenity for the surrounding condominium.

The Chicago Center for Green Technology (CCGT)
The extensive sedum green roof on the CCGT is one of many green features of the LEED platinum rated building. Located on a former brownfield site, the CCGT is a public demonstration, education, and testing site for roofing and sustainable technologies.

COURTESY OF THE CITY OF CHICAGO

COURTESY OF THE CITY OF CHICAGO

Colors change throughout the season around the sunburst pattern. The roof is not open to the public, but limited tours are given to industry professionals, and thousands of people can view the roof from the windows of the 30 nearby buildings that overlook it. Interest in the City Hall roof and in the city's support of green roofing has been so great that a multi-departmental team developed a brochure entitled "A Guide to Rooftop Gardening" to respond to public demand for information.

The city's ongoing temperature monitoring indicates that the green roof is having an effect on the microclimate around City Hall. Roof surface temperatures are consistently lower on the green roof than on the black tar roof of the adjacent Cook County Administration Building. Readings on a 90°F afternoon in August 2001 averaged 100°F on the planted sections of the City Hall roof and 170°F on the surface of the County Building.[2]

Public Policy

In order to promote green roof development in the private sector, the Department of Planning and Development held symposia and focus groups. It found that the lack of green roof data specific to Chicago was an obstacle for many developers. In response, the city constructed an extensive research site on the Chicago Center for Green Technology (CCGT). One of the city-funded heat management research grants helped develop the research site. The CCGT building sits on a former brownfield and has received a platinum LEED rating. The Center houses several green businesses, a training program for disadvantaged adults, and a resource center. Free tours are available, and CCGT hosts seminars on a variety of green technology topics. The building has an extensive sedum green roof and has become a test site for sustainable roofing techniques. Scientists there monitor six different types of green roofs and two reflective roofs, collecting data on temperature differences,

stormwater quality, and stormwater retention. The data will be used both to provide information to the community and to allow city officials to determine optimal roofing techniques for the city.

Private sector green roof development remains a priority for Chicago. In addition to ongoing research and numerous municipal green roofs, in 1999, city heat management grants funded a green roof on the private Schwab Rehabilitation Center, whose roof garden is now part of the hospital's therapy regimen. In late 2002, the city modified plans for new municipal buildings to include green roofs. By adding green roofs to its standards for city projects, Chicago encourages contractors and builders to learn the new technology in order to win contracts.

The Departments of Environment and of Planning and Development have also implemented several policy incentives. The first of these, de-

Peggy Notebaert Nature Museum
Financed by one of the early heat grants, the Peggy Notebaert Nature Museum experiments with a variety of plant palettes, including a wetland area, a small pool, and over eighty sedum, native, and non-native species. A solar panel supplies electricity to a drip irrigation system and a trickle pump, which circulates the water and fills the pool.

Apple Store
The Apple Store on North Michigan Avenue opened in 2003 with an instant green look: a 2,400-square foot sedum roof made of pre-grown modules. Apple installed the roof, which is visible from the class studio/boardroom above, to demonstrate its commitment to improving Chicago's environment.

COURTESY OF THE CONSERVATION DESIGN FORUM

COURTESY OF SHIGEYO HENRIQUEZ

veloped in 2000, was a zoning bonus that allows developers to build at a higher density in the central business district by adding a green roof. The program has been successful because of the high demand for the valuable property in the district—12 developments had used it by 2004. City officials also work with developers, offering tax-increment financing on a case-by-case basis for those who agree to include a green roof.

In 2002, Chicago passed a new energy code, the first municipal ordinance that specifically mentions green roofs. Responding to the city's heat, air, and energy concerns, the energy code adopts widely-used industry standards of energy conservation into the municipal building code.[3] The new regulations mandate a minimum standard of reflectivity for all new roofs, but allows green roofs or solar panels to be used in lieu of reflective roofing materials. The Department of Environment believes that the new requirement will considerably offset the urban heat island effect by decreasing the area of heat-radiating surface in the city. The energy code also creates a distinction between green roofs and rooftop gardens, and clarifies building code regulations for each on egress, fire safety and other concerns. Roof gardens are subject to the stricter regulations applied to spaces intended for human occupancy, but green roofs are exempted. A green roof is subject to the same code as con-

ventional roofs and other spaces not intended for occupancy, enabling developers to implement them at a lower cost.

While Chicago's heat concerns have been the force behind most of its green roof development, the Departments of Environment and of Water Management are also exploring the benefits green roofs provide for stormwater management. In 2001, a series of heavy rainfalls caused flooding, leading to sewer overflow and eventually to beach closures due to E. coli contamination in Lake Michigan. Like the heat wave in 1998, the flooding prompted the city to reevaluate its practices and policies, this time regarding stormwater. Officials are considering granting a utility discount to businesses and residences that use Best Management Practices for wastewater. This incentive would likely include the elimination of sewer connection fees for new structures built with a green roof.

Other Green Building Projects

Green roofs are a critical part of Chicago's greening efforts, one piece of a larger framework of sustainable development for the city. The city has developed new green building guidelines for all new municipal buildings. The Chicago Standard, as it is known, is based on the LEED national standard. While the guidelines were being completed, city officials were training each municipal department in new green techniques

and technologies so that all levels of government would be familiar with the Chicago Standard as soon as it was released.

Chicago has also been working to green various sectors, including housing, energy, and transportation. The city sponsored a green design competition in 2003 and has experimented with renovating its typical low-rise bungalows using paints low in volatile organic compounds and energy-efficient appliances. City officials predict trials of green Section 8 affordable housing in the near future. The city has committed to buying twenty percent of its facility electricity from sustainable sources by 2005. Green fueling stations for police cars are experimenting with different forms of renewable energy and ecological wastewater treatment methods.

Under the leadership of Mayor Richard Daley, the City of Chicago has led the green roof and sustainable urban development movements by example. Chicago currently has more than one million square feet of roof space committed to becoming green, and 20 buildings with green roofs—including a transit building, a fire station, schools, a public health center, a museum, and a number of mid- and low-income residences. Through its strategy of incentives, demonstrations, and policies, the city manifests Mayor Daley's commitment to the environment.

06 | Toronto
A Model for North American Infrastructure Development
By Steven W. Peck

Mountain Equipment Store
Renovations to the company's Toronto store included installation of Toronto's first extensive green roof—designed as a prairie meadow—planted in part with native grasses. The original landscape design included the company's logo planted in sunflowers, but was changed after the sunflower seeds attracted unwanted possums and raccoons. Today ducks nest on the roof.

COURTESY OF MARIE ANNE BOIVIN

Toronto is often called the City of Trees—an appropriate name for a city with three million trees and 19,760 acres of parks, ravines, woodlots, and natural waterfront areas.[1] Despite all its greenery, the city also has its share of environmental problems. Like most major metropolitan areas, Toronto struggles with smog, urban heat island effect, water pollution, and stormwater management. Support for green roofs as a solution to these urban ecological challenges began at the grassroots, but through the work of a broad coalition of businesses, nonprofit organizations, and civic groups, green roofs have become an important priority for Toronto's city government. Toronto's commitment to investigating and implementing an effective multi-functional green roof infrastructure now serves as an inspired template for green roof development in cities across North America.

Progression of Green Roof Movement
The Toronto green roof movement began in 1991 when a volunteer network of horticultur-ists, architects, and community activists founded the Roof Gardens Resource Group. The group developed a resource website, held lectures, and produced fact sheets to promote the benefits of green roofs. The group helped to raise awareness about existing rooftop gardens and proposed a city demonstration project.

In 1998, Canada Mortgage and Housing Corporation funded a research project called *Greenbacks from Green Roofs: Forging a New Industry in Canada* to investigate the benefits of green roofs, explore the potential to establish a profitable new industry, and identify barriers to large-scale implementation. The report was prepared by The Cardinal Group Inc. with input from representatives of the private sector, community groups, and government. The report found significant potential demand for green roofs, with the caveat that the market required stimulation by a set of public policies and incentives to offset the higher capital costs of green roofs over traditional roofing methods.[2] The study also identified a need for better technical information on green roof performance for individual buildings and on a citywide scale.

After the release of *Greenbacks from Green Roofs*, representatives from the business community expressed interest in developing a market for green roof technology. In 1999, six Toronto-based companies[3] formed the Green Roofs for Healthy Cities coalition (GRHC), a membership-based network of public and private organizations whose goal was to create a multi-million dollar market for green roofs. GRHC worked with Toronto city officials to formalize a plan for a city-sponsored green roof demonstration project, and helped to assemble an interdepartmental steering committee—which included the Departments of Public Health, Works, Energy Efficiency, Planning, Heritage, and Facilities—to design and implement it. The committee selected two buildings due for re-roofing—

Queens Quay Roof Garden
Built in the 1980s, the Queen's Quay Roof Garden is early example of the many rooftop gardens that dot Toronto's landscape.

Toronto City Hall Green Roof
The green roof on Toronto City Hall features eight distinct green roof plots, each representing different green roof applications. While most have been successful, the Black Oat Prairie Ecosystem and two food development plots have encountered problems due to soil composition and wind and heat exposure.

COURTESY OF MONICA KUHN

COURTESY OF GREEN ROOFS FOR HEALTHY CITIES

Toronto's City Hall and the Eastview Neighborhood Community Center—and recommended a study of Toronto's Urban Heat Island and green roofs.[4] Around the same time, the city's environmental committee released an Environmental Plan, which advocated green roofs for all new developments. More of a "wish list" than a concrete strategy, the plan was derailed by its proposed multi-billion dollar budget. Nonetheless, the steering committee, the Environmental Plan, and the lobbying efforts helped secure a place for green roofs in the city's Official Land Use Plan. This inclusion, coupled with a successful demonstration project in nearby Chicago, motivated the City Council to approve $280,000 (CDN) for a testing and demonstration project on Toronto City Hall in the spring of 2000.[5]

City-Funded Projects
The 7,000 square foot City Hall roof was the first to be completed in Toronto. It is open to the public and features eight experimental plots, ranging from shallow sedum and alpine perennials to an intensive vegetable garden. In contrast, the inaccessible roof of the Eastview Neighborhood Community Center is an extensive design that serves as a controlled test site. Research on Eastview focuses on stormwater re-

tention, energy efficiency, and impact on the life span of the roof membrane. Both roofs are now part of the Toronto Green Roof Infrastructure Research and Demonstration Project, the parent project that will evaluate the environmental performance research of green roofs. The three-year study and demonstration project is a unique public-private partnership between federal and municipal agencies and local manufacturers, all of whom helped finance the $1.2 million (CDN) demonstration roofs and research. The partnership includes member companies of GRHC, the City of Toronto, the National Research Council's Institute for Research in Construction, Environment Canada, the Toronto Atmospheric Fund, and the Technology Early Action Measures, a federal climate change mitigation program.

In the fall of 2002, the project team released key findings from the Green Roof Infrastructure Technology Demonstration Project's research. The study assumed a modest six percent green roof coverage over 10 years as a baseline figure (representing only one percent of Toronto's total land area), which, it found, could result in an average overall reduction of 1.8°F (1°C) in the urban heat island effect, with a reduction of as much as 3.6°F (2°C) in some areas.[6] An estimated investment of $45.5 million (CDN) per

year for ten years would be required to achieve this level of coverage.[7]

This information provided the data that the research team used to calculate other potential benefits that would result from a minimum of six percent (65 million square feet) green roof coverage. Lower temperatures could reduce greenhouse gases by .72 megatons (MT) and save approximately $1 million (CDN) in energy costs annually, resulting in an additional greenhouse gas reduction of 1.56 MT. Smog alerts would be reduced by five to ten percent and air quality would improve. In addition, a green roof infrastructure would retain 127 million cubic feet of stormwater annually. A stormwater storage tank of the same capacity would cost $60 million (CDN), and offer none of the additional benefits. The study calculated the economic impact of green roofs on jobs, food production, and the development of new economic sectors, and found potential improvements across the board. The study also cited factors more difficult to quantify such as increased biodiversity, additional recreational space, and quality of life.[8]

The study has been instrumental in illustrating the value of government investment in green roof infrastructure development. In March 2004, the

Annex Organics
Until 2001, Annex Organics grew produce in hydroponic containers on the rooftop of the FoodShare distribution warehouse. Specialty crops included tomatillos, cape gooseberries, and ten varieties of hot pepper.

COURTESY OF MONICA KUHN

City Council approved the formation of an Environmental Roundtable to investigate and recommend policies and incentives to encourage green roof construction. The roundtable will present its findings to the city council in January 2005.

Private-Sector Projects

While the city works to establish a municipal green roof framework, some private groups are experimenting with using roof space for agriculture. FoodShare, an organization that helps communities access affordable and healthy food, was the site of a pilot program from 1996 to 2001. Annex Organics used a hybrid hydroponic container system on the roof of the FoodShare warehouse to grow vegetables for sale to organic food stores and high-end restaurants. The Fairmont Hotel chain installed a containerized roof garden on its Royal York Hotel in Toronto to enhance its environmental image. The garden features 17 beds and 23 planter pots planted with herbs and vegetables that are used in the hotel restaurant. The success of the program with the restaurant chefs led the chain to install more substantial plots in San Francisco and Vancouver City—the latter of which has produced $30,000 (CDN) worth of produce in a year.[9] Green roof food production through com-

munity gardens, while a niche application, may have significant benefits for communities where land is scarce and there is a need for community-building infrastructure.

The Green Roofs for Healthy Cities coalition has been very successful at demonstrating the marketability of green roofs to different constituencies—government, private sector, and community groups. Based on the success of the coalition's holistic green roof benefit evaluation in Toronto and research on European cities as well as Portland and Chicago, GRHC has developed a market research and development process to help other North American cities establish green roof programs. The process evaluates a wide range of public and private benefits as the basis for policy support. It has five phases, which guide local leaders as they introduce green roof awareness, engage the community, develop an action plan, gather research data, and finally translate the research into policy options to increase green roof investment. The program recognizes that diversity in climate, architecture, and environment among cities forms the basis for conducting a holistic biophysical and socioeconomic evaluation of public investment in green roof infrastructure. The program is easily adaptable to any locality and has been implemented

in cities such as Vancouver, Calgary, Minneapolis, and Washington.[10]

Flexibility and multidisciplinary participation are the cornerstones of Toronto's green roof movement. Toronto's city government and GRHC have had great success promoting green roofs not for a single issue, but as a solution to a wide range of urban problems. Including many different constituencies has broadened the range of ideas and support for a widespread greening of the city's rooftops, and will provide a basis for the transition to sustainable building practices. It is possible, within a generation, to transform new and existing buildings into engines of renewal, rather than of decay and decline. In the future, restorative, high-performance green buildings will not only take as little from the environment as possible, but will actually generate cleaner water and cleaner air, plus provide a source of renewable energy and enhanced biodiversity. These structures can reconnect city-dwellers to nature, while providing healthy indoor environments for their inhabitants and outstanding operational cost savings for their owners. In order to accomplish this transformation, market mechanisms must be developed to reduce the higher capital costs associated with designing and constructing truly sustainable buildings.

07 | New York City
Greening Gotham's Rooftops
By Colin Cheney

The High Line
Built in the 1930s for freight trains, the High Line fell into disuse with the rise of trucking in the 1950s and gradually became colonized by natural vegetation. In 2003, zoning officials approved a plan to preserve the abandoned elevated rail line as a 1.6-mile long park.

COURTESY OF GEORGE A. FULLER, BUILDER, ARCHIVEOFINDUSTRY.COM   COURTESY OF JOEL STERNFELD

New York City has a long tradition of rooftop plantings, from the 1906 "Babylonian Hanging Gardens" at Coney Island to the renowned gardens atop Rockefeller Center and penthouses around Central Park. These gardens take advantage of the city's great wealth of rooftop area, but are usually private spaces, while New York's public parks and community gardens are limited to the few open spaces on the city's densely-built ground. On the west side of Manhattan, the abandoned High Line trestle railroad has become an example of the potential of publicly beneficial elevated green space—the train track has become overgrown with naturally-occurring plant life and is being reconceived as a park. In 2001, in the midst of New York's ongoing struggles with environmental problems, environmental nonprofit organizations, design professionals, and ecologists began to re-envision vegetated roofspace as ecological infrastructure. With increasing support from government agencies, the movement aims to transform city rooftops into a vibrant network of green roofs to benefit the city at large.

Environmental Challenges
As one of the largest cities in the world, New York City has its share of urban environmental problems. Scientists believe that Manhattan's heat island has existed since the early twentieth century, and estimates of the current urban heat island for New York City range between 3.6°F and 5.4°F above surrounding areas.[1] The Metropolitan East Coast Study, led by Drs. Cynthia Rosenzweig and William Solecki, found that urban temperatures are likely to rise further due to the regional impacts of global climate change. In addition, while the health of the New York/New Jersey estuary in 2003 was the best that it has been since it was first surveyed some one hundred years ago, pollution from stormwater runoff still compromises the health of its waters. Ninety-nine percent of dry-weather discharge from the city's 450 combined sewage overflow pipes has been eliminated, but only sixty-one percent of rainwater is collected and treated.[2] Half of all rainstorms in the city result in combined sewage overflow, pouring an estimated 40 billion gallons of untreated wastewater (rainwater and raw sewage combined) into city waterways every year.[3]

These problems are exacerbated by the city's expanses of pavement. New York City uses almost 2.2 million cubic yards of concrete in making its sidewalks, houses, sewers, subway tunnels, and skyscrapers each year—a volume equivalent to the addition of a new Hoover Dam every 18 months.[4] Like each of the cities in these case studies, an increase in green space would help alleviate New York City's environmental problems. While the city has many parks—including the 820-acre Central Park, the "lungs" of the city—real estate is at such a premium that creation of new open park space is a rare occurrence. Even existing community gardens frequently come under threat from more economically lucrative development plans. Flat rooftops, meanwhile, make up twenty-nine percent of the land area in Manhattan alone. Green roofs on just half of this almost 4,000 acres of roof space would yield green space more than twice the size of Central Park.[5]

Roots of a Movement
Given this expanse of rooftop area, it is not surprising that New Yorkers have long made use of

The Solaire

The Solaire, known as America's first environmentally advanced residential tower, was built in accordance with Battery Park City's strict environmental standards for new construction. The two green roofs—one extensive and inaccessible and one intensive roof deck—were installed to meet requirements for water conservation, site management, and heat island reduction. The building uses a grey water system that recycles rainwater for use in toilets and for irrigation.

COURTESY OF EARTH PLEDGE

roof space for plantings. Pursuit of an ecologically focused green roof network, however, arose from the concerted efforts of a number of individuals. Dr. Paul Mankiewicz, a biologist by training, became interested in green roofs as a way to enrich New York's urban ecology. He worked with the Urban Rooftop Greenhouse Project at the Cathedral of St. John the Divine in the late 1980s to develop a lightweight roof soil of compost and post-consumer polystyrene. Landscape designer Diana Balmori was intrigued by the aesthetic potential of green roofs, an interest that became more pronounced with regard to the rebuilding and re-imagining of the city after the terrorist attacks of September 11, 2001. She designed her first green roof for a Georgian townhouse in Midtown Manhattan.

Earth Pledge, a nonprofit organization focused on sustainable architecture and agriculture, renovated the six-story, 1902 building in 1998 to use as office space and as a showcase for sustainable design. Balmori worked with Earth Pledge executive director Leslie Hoffman to build the green roof. Hoffman recognized the potential of environmen-

tally-focused green roof development, and created the Earth Pledge Green Roofs Initiative in 2001. The program has become the central clearinghouse for green roof design and construction information in New York, and coordinates scientific research, policy development, and public outreach. Seeing the need for both legislative support and scientific research to catalyze green roof technology, the Green Roofs Initiative convened the Green Roofs Policy Task Force and the New York Ecological Infrastructure Study (NYEIS) in 2002.

The Green Roofs Policy Task Force explores options for green roof public policy, and 15 city, state, and federal agencies and offices have taken part. The task force's primary function is to solicit input on the public needs to which green roofs might respond, including housing and social services, public health, environmental quality, zoning and public infrastructure. Task force meetings have detailed the type of information government agencies need, and highlighted areas that need further study. The US Environmental Protection Agency Region 2 (US EPA)

has supported the response to the task force request for more information by funding the informational website GreeningGotham.org. The website has extensive educational materials and features the Green Roof Toolbox, developed for professionals as a central green roof design and construction resource. The site also includes a list of signatories who support the vision of green New York rooftops, including Mayor Michael Bloomberg, Speaker of the New York City Council Gifford Miller, U.S. Senator Hillary Clinton, U.S. Congressman Charles Rangel, many city commissioners and city council members, and other luminaries.

The New York Ecological Infrastructure Study (NYEIS) undertook a comprehensive analysis of the impacts, costs, and benefits of citywide green roofing from 2002 to 2004. Paul Mankiewicz worked with HydroQual, Inc., the environmental engineering firm used by the city to model its sewage system, to investigate stormwater retention benefits of green roofs. NASA researcher Cynthia Rosenzweig and Hunter College's William Solecki,

School of the Future
A student from School of the Future analyzes temperature levels on the school's green roof test plot developed in partnership with NASA and Earth Pledge. The project is part of an effort to develop curriculum centered around green roofs in New York.

Nassau Icehouse Brewery Residence
The rooftops of the Nassau Icehouse Brewery Residence in Crown Heights, Brooklyn, are unlike anything seen in the 1860s, when the original building was constructed. The three roofs, two green and one solar, are part of a larger green renovation of the former brewery's ice storage facility, which also includes radiating heat and use of salvaged materials.

COURTESY OF BIG SUE LLC

COURTESY OF EARTH PLEDGE

leaders of the Metropolitan East Coast study, focused on the urban heat island and the development of standardized protocols for monitoring New York green roofs. The study was designed to provide policymakers with data to quantify how well green roofs address environmental challenges. The research included an ongoing partnership with the New York City Department of Environmental Protection (DEP). As the municipal entity responsible for maintaining the city's sewage system and water quality, DEP has an interest in the outcome of green roof stormwater trials, and has funded two models of green roof performance in Lower Manhattan. The models provide the first significant green roof stormwater analysis using New York City climate and sewage data.

There has been municipal and regional support through various agencies for the first large-scale green roof projects in New York as well. DEP and the US EPA are working with Pace University and Earth Pledge to build a 20,000 square foot green roof at the Lower Manhattan campus of Pace University. The roof covers a city block and is visible

from many surrounding skyscrapers. The Pace green roof has been billed as the city's signature project, and has also received capital construction funding from the offices of the governor and a city councilman. In Queens, the Pratt Institute Center for Community and Environmental Development and Diana Balmori are developing a green roof on a metal-fabrication plant as part of a planned neighborhood-wide test site for green technologies and green industries. The roof will be financed in part by the New York State Energy Research and Development Authority and the building will be monitored for energy use.

Community Support and Action

A number of community groups and housing organizations have installed green roofs as a way to add greenery in limited space. Working with Earth Pledge's Viridian project, six projects have been completed or are under development as of June 2004. Viridian also has two research partnerships with schools—students at the School of the Future monitor stormwater runoff on the school's green roof, and a similar city-sponsored project is

slated for the Bronx. Other city schools, like the Calhoun School, are considering integrating green roof projects into their educational programs.

In the South Bronx, support for general community greening—and green roofs specifically—comes from federal and local officials. The area north of Manhattan has been known for poverty and blight, and is now undergoing a green revitalization. Congressman Jose Serrano granted funding to begin the Bronx River Alliance (BRA), dedicated to cleaning up the Bronx River. Green roofs offer a unique solution for the South Bronx sewershed, whose land area is densely built, with seventy-five percent covered by roofspace.[6] BRA has partnered with Paul Mankiewicz's Gaia Institute and Sustainable South Bronx to incorporate green roofs into the borough's integrated environmental strategies, which also include development of a greenway, wetlands, and soil buffers. In addition, in 2002 the Bronx Borough President, Adolfo Carrion, Jr., and the Bronx Overall Economic Development Corporation established the Bronx Initiative for Energy and the Environment (BIEE). Funded by a one-time appropriation of $1 million, BIEE funds various energy-efficient projects,

Queens Botanical Gardens
The green roof covering the auditorium of the Queens Botanical Garden will be a lush meadow of native species. The green roof, along with geothermal heating, photovoltaic cells, rainwater collection, and a constructed wetland, is part of an effort to design a building to reinforce the organization's educational mission of encouraging environmental sensitivity and celebrating the neighborhood's cultural diversity.

COURTESY OF BKSK ARCHITECTS

including green buildings with solar panels and green roofs.

At the other end of the city, the independent Battery Park City Authority (BPCA) has taken steps towards the creation of green infrastructure in the Battery Park City neighborhood near Ground Zero in Lower Manhattan. The BPCA has mandated stringent environmental requirements for all new buildings, including parameters for site and resource management, water conservation, and energy efficiency. The 27-story Solaire is the first residential high-rise built to these specifications. It consumes thirty-five percent less energy, sixty-five percent less summer electricity, and one-third less water than traditional buildings. The building has 293 apartments and two green roofs—one is accessible to tenants, and the other is inaccessible and was installed solely for its environmental benefits.

Private and Municipal Government Support

A number of private companies have also built green roofs in New York in recent years, including on the Nassau Brewery Icehouse in Brooklyn, an ecologically renovated building with the area's first inaccessible green roof, and the Helena, a luxury high-rise in midtown Manhattan with five extensive green roofs. Some local industries are also exploring the possibilities of greening their facilities. Hugo Neu, the principal recycling facility in New York City and a major property owner with millions of square feet of rooftop, is considering installing green roofs on its buildings.

Through professional education programs and projects such as GreeningGotham.org, green roofs have gained acceptance by the design and building community and the general public. Municipal involvement in the roof at Pace University is a strong indication of the Bloomberg administration's interest in green roofs' potential for New York City and city officials are increasingly interested in green building policies. The City Council is examining green building legislation, including a bill that would require publicly-funded buildings to be LEED silver certified. The Building Department is also seeking to incorporate green building technologies—including green roofs—

into its adoption of the International Building Code. Municipally-funded green roofs are planned on the St. George Ferry Terminal in Staten Island and at the Queens Botanical Gardens.

Advocates of green roof development in New York City believe that success will be marked by the demonstration of green roof benefits to the City's environment, the creation of green roof-supportive local industries, and new green roofs on schools and in low- and moderate-income communities. But true success must also involve a shift in urban consciousness toward more sustainable patterns of living. Jane Jacobs writes that green roofs can be an important part of the re-imagining of the post-industrial city and that "perhaps it will be the city that reawakens our understanding and appreciation of nature, in all its teeming, unpredictable complexity."[7] Through the inspiration and dedication of several individuals and organizations, key players in the New York region are beginning to work toward an infrastructure of green roofs to address environmental problems and rethink the form and meaning of our urban nature.

Green Roof Materials and Components

Composed with the help of Katrin Scholz-Barth and Ed Snodgrass

Waterproofing Membrane
After the PVC membrane is installed, it is tested for leaks with a flood test.

Waterproofing Membrane
The top layer of the modified bitumen membrane is heat welded to the roof. A root barrier will later be applied to prevent penetration of the membrane.

Insulation
Extruded polystyrene is installed over the waterproofing membrane as an insulating layer, reducing energy costs for the building.

COURTESY OF ROOFSCAPES INC.

COURTESY OF BUILDING LOGICS

COURTESY OF ROOFSCAPES INC.

There are seven basic elements to any green roof: waterproofing membrane, root barrier, insulation, drainage layer, filter fabric, growing medium, and plants. These components allow for vegetation to grow on a built surface while protecting the underlying structure.

The modern green roof system was developed in Germany, after years of research and testing. The German Society for Landscape Development and Landscape Design (*Forschungsgesellschaft Landschaftsentwicklung Landschaftsbau* or FLL) has established the most comprehensive material, construction, and maintenance standards for green roof design. FLL guidelines can be ordered directly from www.f-l-l.de, or the American distributor, www.roofscapes.com/FLLguide.htm. A number of manufacturers offer all-in-one systems; easy assembly and warranty make them ideal for large roof surfaces. For small projects, buying the components separately can be cost-effective and allow for increased design flexibility.

With the exception of those that grew unintentionally, each of our building case studies uses these components. Some use complete systems developed by one manufacturer; others are composites designed by the landscape architect or contractor. The materials differ from project to project depending on a variety of factors including function, plantings, geographic region, and manufacturer.

The following appendix outlines some of the material options for a green roof system. This explanation is not intended to be comprehensive; rather it allows the reader to understand the components that make up the green roof case studies. For a more complete explanation of how to build a green roof, see Theodore Osmundson's *Roof Gardens: History, Design, and Construction*, Noel Kingsbury and Nigel Dennett's *Planting Green Roofs and Living Walls*, or the Green Roof Toolbox at www.GreeningGotham.org.

Waterproofing Membrane

The waterproofing membrane safeguards the roof from leakage and therefore is one of the most important elements of any roof—green or not. After the application of the waterproofing membrane, a leak detection test must be performed before applying additional layers. In Germany, roof membranes are tested for compatibility with green roof systems, but in North America, no such standardization exists as of 2004. North American green roof system warranties generally guarantee the integrity of the waterproofing membrane over a certain period of time. In the United States, modified bitumen and PVC membranes most commonly come with these warranties.

Modified bituminous membranes are made by fusing two organic felts with bitumen, a coal byproduct. Synthetic rubber is added to the bitumen for flexibility, elasticity, and strength. Sheets made with Styrene Butadiene Styrene (SBS) are appropriate for green roofs. The membrane is applied by torching down sheets to the roof deck, or spreading the liquid form, which is often referred to as rubberized asphalt.

Thermoplastic membranes, such as PVC, come in long, loose-laid, synthetic sheets. The sheets

are rolled on the deck, overlapping at the joint, and are applied with heat or mechanical fasteners. PVC, more commonly known as vinyl, is a common green roof membrane, and is torched down to the roof deck.

Elastomeric membranes, such as EPDM, are made of synthetic rubber. They are strong and puncture-resistant, but they are not commonly used on green roofs in the United States. EPDM uses adhesive tabs to attach it to the deck, and these tabs are susceptible to root intrusion.

Root Barrier

The root barrier protects the waterproofing membrane and deck from penetration by aggressive roots. Membranes with adhesive tabs (e.g., EPDM) and those made with organic materials on which plants like to feed (e.g., modified bitumen) are particularly susceptible to root invasion. Those made of synthetic materials (e.g., PVC) are naturally root-repellant and do not need an extra root barrier.

A polyethylene sheet serves as an effective root barrier because it is made of synthetic plastics.

Some companies offer waterproofing membranes or filter fabric inlaid with *copper foil* or *copper hydroxide*, as copper is a natural root repellant.

Insulation

Insulation is not a structurally necessary component of a green roof, but most building codes require it in standard roof construction to prevent heat loss. A green roof alone minimizes energy use in the summer but is not an effective insula-

Drainage Layer

Good drainage is essential for a successful green roof. A lightweight synthetic drainage board ensures well-drained soil and proper aeration of the roots.

COURTESY OF ROOFSCAPES INC.

Substrate

Green roof growing medium—known as substrate—is typically a mix of lightweight mineral aggregate and organic matter. Here it is installed with a blower truck.

COURTESY OF ZINCO GMBH

Vegetation

Pre-grown sedum mats provide instant vegetation coverage. Other planting methods for extensive green roofs include plugs, cuttings, and seeds, which take one to three years to become fully established.

COURTESY OF BUILDING LOGICS

tor in the winter. An additional insulation layer maximizes energy savings by reducing heat and air conditioning use. Insulation can be applied in several ways. In regular roof construction, the insulation is either placed beneath the roof deck or between the waterproofing membrane and roof deck. Either option is suitable for a green roof. The most common insulation installation method for a green roof is the inverted roof membrane assembly (IRMA), which places the insulation above the waterproofing membrane. In IRMA application, the insulation protects the membrane, and can be salvaged in the case of re-roofing.

Polyisocyanurate, also known as isoboard, is a common insulator for below the waterproofing membrane.

Extruded polystyrene is the only type of insulation approved for IRMA application, and has high water resistance and compressive strength.

Drainage and Retention Layer

Good drainage is the key to successful rooftop plant propagation. The drainage layer prevents oversaturation, ensures that roots are ventilated, and provides roots with extra space to grow. Many drainage layers also help retain water or are partnered with retention mats. Water flows naturally off pitched roofs (those with a slope larger than 5°) making a drainage layer unnecessary except to aid with extra retention. Flat roofs require a drainage layer to direct water off the roof and prevent puddling.

*Synthetic drainage board*s are usually made of strong, lightweight plastic and come in a num-

ber of shapes (e.g., egg cartons, honeycombs, wire plastic sheets). They are often made with crevices and/or attached absorbent mats where water can be stored in times of drought.

Granular aggregate is made of a mineral mixture, such as clay, lava, expanded slate, slag, brick, or foamed glass. This kind of base has been used as drainage for centuries, and is often made from the primary components in the growing medium. It is heavier than synthetic drainage mats but stores water more effectively.

Retention layers are made of felt or other absorbent recycled materials that will store water for plants to use in times of drought.

Filter Fabric

A *geotextile filter fabric* must be placed between the drainage layer and the growing medium to keep the substrate in place. It is usually made of polyester or non-woven polypropylene.

Growing Medium

Although it is the top layer, the growing medium is the foundation of the green roof, providing the nutrient base and space for the plants to grow. Humus and topsoil are heavy and do not allow for the necessary drainage, aeration, and replacing of organic material necessary on a green roof. The growing medium for a rooftop is made from different components than ground soil—a mineral base with minimal organic material—and is therefore often referred to as *substrate*. Green roof substrate composition is determined by water retention capacity, weight, aeration, and nutrient retention, which are based on porosity

and grain size.

Coarse soils have the lowest water retention capacity but allow the greatest root aeration—necessary because roots can rot without sufficient air. Fine soils more effectively absorb and hold water and nutrients in reserve for dry periods. A mixture of grain sizes retains the most water and nutrients and allows for aeration of the roots. A good green roof substrate is often a mix of a lightweight aggregate and organic matter. Low-weight, high-porosity aggregates like expanded shale and clay are particularly suited for rooftops and have stable grains that will not get windblown. Other common materials are expanded clay, expanded shale, crushed brick, lava, and volcanic glass.

While manufactured lightweight mixes are most common for green roofs, some projects oriented to plant restoration or habitat recreation use onsite materials. Swiss and British rubble roofs use debris from brownfield sites to recreate habitat of local invertebrates and bird species.

Plant Selection

A rooftop is an unusual environment for plant life; not all species will thrive in its particular conditions. The selection of appropriate plants is essential to both the aesthetic and environmental function of the green roof. Studies in Germany have shown that green roofs retain stormwater and insulate most effectively when they are full-grown and flourishing. Selecting plants appropriate for rooftop conditions requires consideration of certain microclimatic and individual plant characteristics that may not be relevant to ground gardens.

Sedum 'Green Spruce'

Jovibarba hirta

Sedum 'Sichotense'

Sedum cauticola 'Lidakense'

Sedum kamtschaticum

Sedum 'Bertram Anderson'

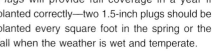

COURTESY OF EMORY KNOLL FARMS

Climate and Microclimate: As in any landscaping project, green roof plants must be appropriate to the climate. Hardiness zone charts can help determine which plants are suited to the region, as determined the average low temperature. Lows in Zone 1 range from -45°F to -50°F, and in Zone 10 average 30°F to 40°F. Rooftops have microclimates as well—specific sun, shade, and wind patterns that affect plant growth. High walls or other irregular features can create wind tunnels or shady areas.

Weight: The weight of plants at maturity is important to long-term load concerns. Low-lying sedums rarely have a biomass of more than 2.5 pounds per square foot, while plants that add mass through vertical growth, like some cactuses, or those that create topsoil through their life cycle, like some grasses, can affect the long-term weight load on the roof.

Seeding plants: Some plants produce easily airborne seeds, which should be considered. They can create unintended outcroppings on the roof itself or in nearby ground gardens, or disturb the surrounding ecosystem. Plant seeds are also a way to attract wildlife, whether intended or unintended.

Longevity: A basic, low-maintenance green roof is most easily planted with perennials and re-rooting plants like sedums. Intensive roof gardens can support annuals and higher-maintenance plants. Some native species, species endemic to a particular area, have been shown to be successful on rooftops as well.

Sedums, a variety of drought tolerant succulent, have become the most common extensive green roof plants as they fulfill most green roof environmental and design considerations. They have a wide range of applicability for hardiness and heat zones. They have a low weight at maturity and re-root quickly, so they can be applied in cuttings. They are non-invasive, drought-resistant, and come in a wide range of colors from blood reds to evergreens. Drought-tolerant succulents store water efficiently by transpiring only at night, when water loss is minimal. Once established, they can survive on rainwater alone and withstand high temperatures.

Planting Types

For extensive green roofs, there are a number of planting methods, which vary by cost and coverage. For immediate full coverage, the methods are *pre-cultivated vegetated mats* (or *carpets*) and *modular systems. Grow-in* techniques do not offer immediate coverage, but are up to five times less expensive. Grow-in techniques include plugs, cuttings, and seeds. Many green roof experts recommend plugs as the best value of grow-in techniques.

Pre-cultivated mats are long rolls of pre-grown vegetation, usually sedums, which are laid down and attached to the roof.

Modular systems are perforated interlocking blocks containing drainage materials, soil, and pre-grown plants. Similar modules are often used on football fields and can be dismantled easily.

Plugs will provide full coverage in a year if planted correctly—two 1.5-inch plugs should be planted every square foot in the spring or the fall when the weather is wet and temperate.

Cuttings are more sensitive to climate variance, and need consistent moisture. If rainfall is infrequent, irrigation and a blanket of mulch may be necessary.

Seeds are the least expensive option when applied properly, but can take up to three years to be fully established.

Planting methods for intensive green roof gardens are similar as in ground gardens. Seeds, bulbs, cuttings, and root balls can all be used. Large plants should be planted with a contained root ball to safeguard against root penetrations and to provide a more immediate visual amenity.

A green roof is best planted in the spring or fall when the temperature is temperate, and plants should always be delivered as close to the day of planting as possible to prevent wilting. Unless the climate is very wet, additional irrigation is necessary for the first few months as the plants acclimate to new conditions. Once they are established, extensive roofs require very little maintenance. Plantings can benefit from a slow-release organic fertilizer applied once a year for the first five years to fully establish the roots. In addition, selective weeding during the initial period is important to make sure no competitive species begin to dominate. After the first five years, extensive green roofs can be left alone, other than an annual visit to check drains and troubleshoot.

Design Details

01 | Glen Patterson's Garden on the Escala

BUILDING TYPE
Residential Multi
NEW/RETROFIT
New
GREEN ROOF TYPE
Intensive
GREEN ROOF SIZE
2,000 s.f.
PERCENTAGE OF ROOF COVERED
100%

Green Roof System Details

SYSTEM MANUFACTURER: n/a
MEMBRANE: Root repellant membrane
DRAINAGE: .25-in. filter drain mat, 1-in. insulation, 3-in. clear crushed granular drainage material, and a geotextile filter fabric. Secondary roof membrane and an additional filter drain mat topped by 2-in. concrete in some areas.
SOIL MEDIUM: Specially-designed soil mix containing black pumice pebbles; graded sand; coconut fiber for organic content; and zeolite to improve the action-exchange ratio.
SOIL DEPTH: 12-24 in. The entire garden is 12 in. deep, with areas of 24 in. for deep-rooting trees.
PLANTS:
TREES: Nothofagus antartica 'Antarctic Beech'; Pinus parviflora 'Japanese White Pine'; Juniperus chinensis 'Chinese Junipers'; Chamaecyparis pisifera 'filifera' 'Threadneedle Cypress'; Chamaecyparis pisifera 'Squarrosa Snow'; Cryptomeria japonica 'Sekkan-suggi'; Rhododendron macabeanum; Rhododendron oreotrephes; Pinus thunbergii 'Japanese black pine'; Quercus phillyreoides 'Japanese evergreen oak'; Sequoia sempervirens 'Prostasta'/'Cantab'; Cupressus glabra 'Smooth Arizona Cypress'; Acer p.d. 'Shishigashira,' or 'Lion's head Maple'; Acer palmatum 'Dissectum Atropurpureum' (This Japanese maple is nearly 100 years old, and is planted over a column for structural support.)
SMALLER TREES AND SHRUBS: Ginkgo bilobas including 'Variegata' and 'Pendula'; Acer p. 'Senkaki'

'Coral Bark Japanese Maple'; Acer shirishawanum 'aureum' 'Japanese Moon Maple'; Continus hybrid 'Grace'; C.coggygria 'Velvet Cloak'; X C. obovatus; Tsuga mertensiana 'Mountain hemlock'; Salix magnifica (a willow with magnolia-like foliage); Cupressus cashmeriana conifer; Podocaarpus totara—a podocarp from New Zealand; Lagarostrobus franklinii (a conifer from Tasmania called 'Huon Pine'); various Cryptomeria japonicas including 'vilmoriniana,' 'Dacrydioides,' and 'Nana'; Zenobia pulverulenta; C. Dwarf Rhododendrons (all species: R. albrechtii, calostrotum 'Gigha,' campylogunum, 'myrtilloides,' cephalanthum, fastigiatum, ferrugineum, flavidum, fletcheranum, globigeron, hypenanthum, impeditum, keiskei, keleticum, lepidostylum, kiusianum, macrosepalum 'linearifolium,' megeratum, pachysanthum, prostatum, quinquefolium, roxianum oreonastes, sargentianum, serpyllifolium tsariense,) and tender Rhododendron species including R. edgeworthii and R. lyi.
GROUNDCOVERS: Various species of vacciniums and gaultherias, Cornus Canadensis, various species of cyclamen asarum, small ferns, and ophiopogon Linnaea borealis.
ROCK GARDEN: 20 large pieces of tufa were installed for alpine and rock plants. The plants include several species of Sargentianum Androsace, Lewisia, and Daphne arbuscula.
PERENNIALS: Early bloomers include various trillium species, sanguinaria epimediums, dodecatheons, shortia, uniflora, and certain primulas. Later the arisaemas and meconompsis provide exciting floral displays. Some tender plants have survived the winters, perhaps due to heat from the condominium below. These include Pachystegia insignis and Myosotidium hortensa, both from New Zealand, and various beautiful forms of Aeonium, such as tabuliiforma and a black leaf hybrid called 'Zwartkop.'
PLANTING METHOD: Trees were cloud-pruned to a 3-ft. diameter and 1.5-ft. depth, bound, and replanted in a ground garden for one year to acclimate roots to the trimming. All trees were successfully planted at the end of the year.
SPECIAL FEATURES: The garden has an automatic irrigation system; three ponds (30-in., 18-in. and 12-in. depth); a waterfall; and artificial rocks made of steel mesh and blown cement mix sprayed with color for pebbling effect.
GREEN BUILDING FEATURES: n/a
COST: n/a
WEIGHT: Roof strengthened to load-bearing of 250lbs.
SUBMITTED BY: Glen Patterson

02 | Casa Bauträger

BUILDING TYPE
Residential
NEW/RETROFIT
New
GREEN ROOF TYPE
Intensive
GREEN ROOF SIZE
4,295 s.f.
PERCENTAGE OF ROOF COVERED
100%

Green Roof System Details

SYSTEM MANUFACTURER: Optigrün
MEMBRANE: PVC
DRAINAGE: Optigrün expanded clay
SOIL MEDIUM: Optigrün intensive substrate
SOIL DEPTH: 15.75 in.
PLANTS: n/a
PLANTING METHOD: Various
SPECIAL FEATURES: Winter garden and pond
GREEN BUILDING FEATURES: n/a
COST: n/a
WEIGHT: 102 lbs/s.f.
SUBMITTED BY: Edmund Maurer, City of Linz

03 | KPMG

BUILDING TYPE
Office
NEW/RETROFIT
Retrofit

GREEN ROOF TYPE
Intensive
GREEN ROOF SIZE
44,100 s.f.
PERCENTAGE OF ROOF COVERED
67%

Green Roof System Details

SYSTEM MANUFACTURER: n/a
MEMBRANE: Modified Bitumen and PVC
DRAINAGE: 3.5-in. gravel drainage
SOIL MEDIUM: Topsoil for standard areas; lava mixed with zeolith for wetlands.
SOIL DEPTH: 12-17 in.
PLANTS: Standard species for ornamental plantings; trees, shrubs, ground cover, and wetland vegetation such as carex and iris. The principle plantings are a patented type of riparian vegetation called logatainer and phytolyse.
PLANTING METHOD: Planted by hand
SPECIAL FEATURES: Water cycled by pump. The plant roots naturally purify the water in the creek and swamp zone.
GREEN BUILDING FEATURES: Artificial wetland and rainwater collection
COST: $100/s.f.
WEIGHT: n/a
SUBMITTED BY: Ulrich Zens

04 | Heinz 57 Center/Gimbel's Building Rehabilitation

BUILDING TYPE
Office
NEW/RETROFIT
Retrofit
GREEN ROOF TYPE
Extensive
GREEN ROOF SIZE
12,000 s.f.
PERCENTAGE OF ROOF COVERED
100% at penthouse level, 33% of total roof area

Green Roof System Details

SYSTEM MANUFACTURER: Roofscapes Inc.
MEMBRANE: EPDM
DRAINAGE: Granular drainage zone and a low-profile perforated conduit. Water retention is based on the German FLL standards: efficient media water absorption—30% by volume maximum water capacity; six roof drains 1/2000 s.f.; maximum water capacity at an estimated 1.75 inches fully drained. Predicted annual rainfall retention: 55%.
SOIL MEDIUM: 90% mineral; 10% organic
SOIL DEPTH: 5 in. The system uses a 2-in. base for drainage and 3-in. upper layer.
PLANTS: 50% sedum and 50% meadow perennials.
PLANTING METHOD: 72-cell plugs hand planted on 12-in. centers; direct seeding.
SPECIAL FEATURES: Non-irrigated integration of decks, patio areas, and walkways.
GREEN BUILDING FEATURES: n/a
COST: $16/s.f. including initial maintenance
WEIGHT: 30 lbs/s.f.
SUBMITTED BY: Charlie Miller, Roofscapes Inc.

05 | North German State Clearing Bank (NORD LB0)

BUILDING TYPE
Office
NEW/RETROFIT
New
GREEN ROOF TYPE
Extensive and intensive
GREEN ROOF SIZE
84,584 s.f.
PERCENTAGE OF ROOF COVERED
60%

Green Roof System Details

SYSTEM MANUFACTURER: ZinCo GmbH
MEMBRANE: Wolfin System
DRAINAGE: ZinCo FD 40
SOIL MEDIUM: 80% crushed brick; 20% compost
SOIL DEPTH: 4.8-8 in.
PLANTS: Sedums and 30,000 bulbs, perennials, and herbs.
PLANTING METHOD: Sedum cuttings and bulbs planted by hand.
SPECIAL FEATURES: n/a
GREEN BUILDING FEATURES: Window ventilation, geothermal chilling, daylight redirection, and fuel cells.
COST: $23.06/s.f.
WEIGHT: n/a
SUBMITTED BY: Taylor & Company

06 | A&C Systems

BUILDING TYPE
Office and storage
NEW/RETROFIT
New
GREEN ROOF TYPE
Extensive
GREEN ROOF SIZE
3,229.2 s.f.
PERCENTAGE OF ROOF COVERED
30%

Green Roof System Details

SYSTEM MANUFACTURER: Floradak
MEMBRANE: Rubberized asphalt
DRAINAGE: Floradak Green Roof System
SOIL MEDIUM: Floradak extensive soil
SOIL DEPTH: 4.3 in.
PLANTS: Sedums
PLANTING METHOD: n/a
SPECIAL FEATURES: Grey water system
GREEN BUILDING FEATURES: Wind and passive solar power
COST: $4-5/s.f.
WEIGHT: 19.5-33.8 lbs/s.f.
SUBMITTED BY: Mark Depreeuw, Architecten Atelier Mark Depreeuw

07 | Sechelt Justice Services Centre

BUILDING TYPE
Institutional
NEW/RETROFIT
New
GREEN ROOF TYPE
Extensive
GREEN ROOF SIZE
5,000 s.f.
PERCENTAGE OF ROOF COVERED
40%

Green Roof System Details

SYSTEM MANUFACTURER: Soprema

MEMBRANE: Modified Bitumen

DRAINAGE: Miradrain 9800

SOIL MEDIUM: 60% black pumice and 40% soil amendment

SOIL DEPTH: 3 in.

PLANTS: *Native Species:* Sedum spathuli folium, sedum douglas glauca, sedum spurium tricolor, festuca ovina, festuca viridula. *Non-Native Species:* Broad leaved stonecrop, douglas sedum, european dragon's blood, rocky mountain, fescue green leaf, fescue poa arctica, artic bluegrass.

PLANTING METHOD: 1 in. diameter plugs from 72 trays were all hand planted.

SPECIAL FEATURES: Temporary irrigation system shuts off after 10 months.

GREEN BUILDING FEATURES: Passive solar, earth duct system that circulates cool air from an underground pipe. A weather station controls opening upper windows. Rainwater is collected from a chute flowing into an infiltration basin/stream.

COST: $7/s.f.

WEIGHT: 15lbs/s.f.

SUBMITTED BY: Randall Sharp, Sharp & Diamond Landscape Architecture & Planning

08 | Hill House

BUILDING TYPE
Residential, single family.

NEW/RETROFIT
New

GREEN ROOF TYPE
Extensive

GREEN ROOF SIZE
2,500 s.f.

PERCENTAGE OF ROOF COVERED
100%

Green Roof System Details

SYSTEM MANUFACTURER: n/a

MEMBRANE: Standard hot-mopped

DRAINAGE: Drain tiles built into roof with extra gravel between membrane and soil.

SOIL MEDIUM: 50% road base and 50% topsoil

SOIL DEPTH: 8 in.

PLANTS: Winter rye, clover, wildflowers

PLANTING METHOD: Seeded with winter rye

SPECIAL FEATURES: Rainbird sprinklers

GREEN BUILDING FEATURES: Passive solar, solar hot water, and grey water system

COST: $100/s.f.

WEIGHT: 90 lbs/s.f.

SUBMITTED BY: Steve Badanes, Jersey Devil Design/Build

09 | Nine Houses

BUILDING TYPE
Residential, single family

NEW/RETROFIT
New

GREEN ROOF TYPE
Intensive

GREEN ROOF SIZE
37,674 s.f.

PERCENTAGE OF ROOF COVERED
100%

Green Roof System Details

SYSTEM MANUFACTURER: n/a

MEMBRANE: Modified Bitumen

DRAINAGE: No drainage mat. Insulation is 9.8-in. layer of foam made of recycled glass, with filter fabric on top.

SOIL MEDIUM: Landfill; humus

SOIL DEPTH: 2.3-6.5 ft.

PLANTS: n/a

PLANTING METHOD: Seeded

SPECIAL FEATURES: Foam made of recycled glass serves as insulation.

GREEN BUILDING FEATURES: Heat pump, spray concrete, underground construction.

COST: n/a

WEIGHT: n/a

SUBMITTED BY: Peter Vetsch

10 | Life Expression Chiropractic Center

BUILDING TYPE
Office

NEW/RETROFIT
New

GREEN ROOF TYPE
Extensive

GREEN ROOF SIZE
6,000 s.f.

PERCENTAGE OF ROOF COVERED
100%

Green Roof System Details

SYSTEM MANUFACTURER: Roofscapes Inc. and Sarnafil

MEMBRANE: PVC

DRAINAGE: Specialized media absorb and retain rainfall while remaining fully drained. Based on the German FLL guidelines, the system's maximum water capacity is 1.75 in., with 55% annual rainfall retention.

SOIL MEDIUM: 90% mineral; 10% organic.

SOIL DEPTH: 5 in.

PLANTS: Sedum

PLANTING METHOD: Plugs are planted by hand on 12-inch centers.

SPECIAL FEATURES: 25% of the roof has a slope of 14 degrees; 58% slopes at 30 degrees. The media is stabilized using roof battens, slope restraint panels, and reinforcing mesh. To protect the roof from wind erosion until the plants were established, the media surface was covered with a photodegradable wind blanket mesh, which has since disappeared into the cover vegetation. Winner of the 2004 Green Roof Awards of Excellence.

GREEN BUILDING FEATURES: Floor radiant heating, partial earth sheltered structure; daylighting through a cupola clerestory.

COST: $7/s.f.

WEIGHT: 28 lbs/s.f.

SUBMITTED BY: Charlie Miller, Roofscapes Inc.

11 | Arbroath Abbey Visitor Center

BUILDING TYPE
Institutional

NEW/RETROFIT
New

GREEN ROOF TYPE
Extensive

GREEN ROOF SIZE
1,137 s.f.

PERCENTAGE OF ROOF COVERED
70%

Green Roof System Details

SYSTEM MANUFACTURER: Erisco Bauder

MEMBRANE: Modified Bitumen

DRAINAGE: Water retention fleece

SOIL MEDIUM: Recycled crushed brick and expanded shale

SOIL DEPTH: 1 in.

PLANTS: Sedums

PLANTING METHOD: Pre-cultivated sedum mats

SPECIAL FEATURES: n/a

GREEN BUILDING FEATURES: Passive solar; thermal mass; use of natural materials such as untreated timber; natural ventilation.

COST: n/a

WEIGHT: n/a

SUBMITTED BY: Simpson & Brown Architects

12 | Mashantucket Pequot Museum and Research Center

BUILDING TYPE
Institutional

NEW/RETROFIT
New

GREEN ROOF TYPE
Intensive

GREEN ROOF SIZE
52,000 s.f.

PERCENTAGE OF ROOF COVERED
35%

Green Roof System Details

SYSTEM MANUFACTURER: American Hydrotech

MEMBRANE: Rubberized asphalt

DRAINAGE: Protection course/root barrier; 5-in. extruded polystyrene insulation; drainage mat; water retention mat; water retention reservoir; filter fabric.

SOIL MEDIUM: Topsoil sand and organic material

SOIL DEPTH: 9-36 in. In most areas, the depth is 9-14 in. Depth reaches 36 in. at localized tree pits.

PLANTS: *Primary Varieties:* Lowbrush Blueberry; Willowleaf Cotoneaster; Periwinkle; Regosa Rose. *Secondary Varieties:* Buglewood Alium; Little Bluestem; Artemisia; New England Astor; False Indigo; Willowleaf Cotoneaster; Seakale; Purple Cone Flower; Globe Thistle; Joe-pye Wood; Baltic Ivy; Daylily; English Lavender; Grayfeather; Wild Bergamot; Switch Grass; Virginia Creeper; Boston Ivy; Rugosa Rose; American Burnet Pincushion Flower; Sedum Goldenrod; Aaron's-rod; Lowbush; Mullein; Veronica; Sod.

TREES: Red Maple; Japanese Tree; Lilac.

HERB GARDEN: 46 varieties

PLANTING METHOD: Planted by hand

SPECIAL FEATURES: Electronically controlled drip for planters; sprinklers for turf.

GREEN BUILDING FEATURES: Extensive use of sheeting and other measures to reduce the area of site disturbance; infiltration pits to eliminate stormwater discharge to adjacent sacred wetlands; insulated building envelope; passive solar devices including sunscreens and high performance glazing with ceramic frits and low-e coatings; occupancy sensors to reduce lighting costs; heat recovery units; recyclable or high recycled content building materials; low VOC targets for all building materials.

COST: n/a

WEIGHT: 100 lbs/s.f. live roof load

SUBMITTED BY: Susan T. Rodriguez, Polshek Partnership Architects

13 | Schiphol Plaza

BUILDING TYPE
Transportation hub

NEW/RETROFIT
New

GREEN ROOF TYPE
Extensive

GREEN ROOF SIZE
87,188.4 s.f.

PERCENTAGE OF ROOF COVERED
90%

Green Roof System Details

SYSTEM MANUFACTURER: Xeroflor

MEMBRANE: Plastic

DRAINAGE: Xeroflor XF107: mineral wool

SOIL MEDIUM: n/a

SOIL DEPTH: 1.3 in.

PLANTS: Xeroflor XF301 moss sedum pre-cultivated mat

PLANTING METHOD: Pre-cultivated mat

SPECIAL FEATURES: n/a

GREEN BUILDING FEATURES: n/a

COST: $4/s.f.

WEIGHT: 9.2 lbs/s.f.

SUBMITTED BY: Jan Benthem, Benthem Crouwel Architects

14 | Primary and Secondary School

BUILDING TYPE
Institutional

NEW/RETROFIT
Retrofit

GREEN ROOF TYPE
Extensive

GREEN ROOF SIZE
15,000 s.f.

PERCENTAGE OF ROOF COVERED
100%

Green Roof System Details

SYSTEM MANUFACTURER: ZinCo GmbH

MEMBRANE: Bitumen

DRAINAGE: Protection and retention mat SSM 45 Floradrain® FD 25 in flat areas; Floratec FS 75 in sloped areas. Filter Sheet SF.

SOIL MEDIUM: ZinCo System soil "Sedum Carpet"

SOIL DEPTH: 3 in.

PLANTS: ZinCo System vegetation "Sedum Carpet"

PLANTING METHOD: Sedum cuttings and pre-cultivated sedum mats

SPECIAL FEATURES: The PV panels from the Solar-Fabrik AG company are held down by the weight of the green roof only. There was no additional penetration through the waterproofing to fix the panels. This system is available for new and retrofit green roofs.

GREEN BUILDING FEATURES: Active and passive solar.

COST: $4/s.f.

WEIGHT: 18 lbs/s.f.

SUBMITTED BY: ZinCo GmbH

15 | Beddington Zero Emission Development (BedZED)

BUILDING TYPE
Residential, multi-family

NEW/RETROFIT
New
GREEN ROOF TYPE
Extensive
GREEN ROOF SIZE
333,519 s.f.
PERCENTAGE OF ROOF COVERED
n/a

Green Roof System Details
SYSTEM MANUFACTURER: RAM Roof Garden Consultancy Ltd
MEMBRANE: Modified Bitumen
DRAINAGE: Synthetic drainage layer; 11.8-in. insulation
SOIL MEDIUM: Topsoil
SOIL DEPTH: 1-11.8 in. The deeper areas are accessible.
PLANTS: Sedums and some small trees.
PLANTING METHOD: Sedum mat for inaccessible; plugs for accessible.
SPECIAL FEATURES: Porous pipe irrigation uses water from the onsite sewage treatment.
GREEN BUILDING FEATURES: High occupancy; natural lighting; local building materials; passive solar.
COST: n/a
WEIGHT: n/a
SUBMITTED BY: Mathew Frith, Peabody Trust

16 | EcoHouse

BUILDING TYPE
Residential, multi-family
NEW/RETROFIT
Retrofit
GREEN ROOF TYPE
Extensive and intensive
GREEN ROOF SIZE
18,300 s.f.
PERCENTAGE OF ROOF COVERED
50%

Green Roof System Details
SYSTEM MANUFACTURER: n/a
MEMBRANE: Modified Bitumen
DRAINAGE: n/a
SOIL MEDIUM: Turf with vermicompost
SOIL DEPTH: 1.57-3.15 in.
PLANTS: *Vegetables:* Broccoli; cucumbers; dill; lettuce; parsley; tomatoes; zucchini. *Berries:* Currants; gooseberries; strawberries. *Seedlings:* Apple tree; currants; flowers; lawn grass.

PLANTING METHOD: Hand planted in pen beds; low beds covered with plastic; greenhouses.
SPECIAL FEATURES: Greenhouses installed between ventilation shafts.
GREEN BUILDING FEATURES: n/a
COST: .60/s.f.
WEIGHT: n/a
SUBMITTED BY: Valentin Yemelin, GRID ARENDAL

17 | St. Luke's Science Center Healing Garden

BUILDING TYPE
Institutional
NEW/RETROFIT
Retrofit
GREEN ROOF TYPE
Intensive
GREEN ROOF SIZE
15,500 s.f.
PERCENTAGE OF ROOF COVERED
56.7%

Green Roof System Details
SYSTEM MANUFACTURER: n/a
MEMBRANE: PVC
DRAINAGE: Layer of obsidian perlite with a porous PVC pipe
SOIL MEDIUM: 30% perlite; 70% topsoil
SOIL DEPTH: 19.7-27.6 in.
PLANTS: Ilex pedunculosa; sophora japonica; lagerstroemia indica; cornus officinalis; malus halliana; styrax japonica; Ilex serrata; hydrangea macrophylla form; macrophylla; lespedeza bicolor var japonica; weigela coraeensis; pyracantha cv; viburnum dilatatum; euonymusalatus; callicarpa japonica; ligustrum japonicum; rhododendron spp; rhododendron indicum; gardenia jasminoides; hypericum patalum; corylopsis spicata; spiraea thubergii; rhaphiolepis umbellate; ilex crenata cv; deutzia crenata; enkianthus perulatus; abelia grandiflora; spiraea japonica; lampranthus cv; zephyranthes cv; agapanthus africanus; leucothoe fontanesiana; reineckea carnea; farfugium japonicum; hosta cv; ophipogon japnicus; ajuga reptans; pachysandra terminalis; hypericum calycinum; liriope muiscari; bletilla striata; vincca mino; phlox subulata; stauntonia hexaphylla; rosa cv; iris japonica; hedera cv.
PLANTING METHOD: Hand planted. Trees limited to 6.5 ft.
SPECIAL FEATURES: 4.5 in.-wide gravel drainage path around the edge to prevent leaks.
GREEN BUILDING FEATURES: n/a

COST: $63/s.f.
WEIGHT: 102 lbs/s.f.
SUBMITTED BY: St. Luke's International Hospital

18 | Chicago City Hall

BUILDING TYPE
Institutional
NEW/RETROFIT
Retrofit
GREEN ROOF TYPE
Extensive and intensive
GREEN ROOF SIZE
22,000 s.f.
PERCENTAGE OF ROOF COVERED
56%

Green Roof System Details
SYSTEM MANUFACTURER: Roofscapes Inc., Optima System, and Sarnafil
MEMBRANE: PVC
DRAINAGE: 2-8 in. layer of light-weight gravel drainage medium
SOIL MEDIUM: Optima growth media mixed at Midwest Trading
SOIL DEPTH: 4 in., 6 in., 18 in.
PLANTS: 20,000 plants of more than 150 varieties including 100 shrubs, 40 vines, and 2 trees, arranged in a starburst pattern. For a full list, please see the following URL: www.cityofchicago.org.
PLANTING METHOD: Planted by hand
SPECIAL FEATURES: Drip irrigation, rainwater catchment, and recycling.
GREEN BUILDING FEATURES: n/a
COST: $45.50/s.f.
WEIGHT: Saturated weight for 4 in. area is 30 lbs/s.f.; 60lbs/s.f. for 6 in.; 90lbs/s.f. for 18 in.
SUBMITTED BY: Conservation Design Forum

19 | ACROS Fukuoka

BUILDING TYPE
Office

NEW/RETROFIT
New
GREEN ROOF TYPE
Intensive
GREEN ROOF SIZE
100,000 s.f.
PERCENTAGE OF ROOF COVERED
80%

Green Roof System Details

SYSTEM MANUFACTURER: Katamura Tekko Co.
MEMBRANE: Rubber membrane
DRAINAGE: Collection reservoirs and filtering for irrigation overflow.
SOIL MEDIUM: Specific to plant selection.
SOIL DEPTH: 12-24 in.
PLANTS: Japanese indigenous plants were primarily used. In some areas certain non-native annuals or perennials were selected for their flowering season and color. Plant selection was based on hardiness and evergreen qualities. Several types of grasses and perennial flowers were also used to enhance aesthetics.
PLANTING METHOD: The terraced gardens are built vertically. Height progression begins with low-lying hanging plants, grasses, ground cover plants, and moves upwards with flowering shrubs, branching shrubs, miniature trees, and dwarf trees and shrubs.
SPECIAL FEATURES: Rainwater catchment with sensors for irrigation and water reuse.
GREEN BUILDING FEATURES: n/a
COST: n/a
WEIGHT: n/a
SUBMITTED BY: Emilio Ambasz & Associates

20 | Church of Jesus Christ of Latter Day Saints

BUILDING TYPE
Institutional
NEW/RETROFIT
New
GREEN ROOF TYPE
Extensive and intensive
GREEN ROOF SIZE
174,240 s.f.
PERCENTAGE OF ROOF COVERED
Over 50%

Green Roof System Details

SYSTEM MANUFACTURER: American Hydrotech
MEMBRANE: Rubberized asphalt

DRAINAGE: American Hydrotech drainage mat; mirafi geotextile filter fabric
SOIL MEDIUM: Expanded aggregate; organic matter
SOIL DEPTH: 2-8 in.
PLANTS: The entire green roof uses native plant materials. The intensive garden is planted with a variety of native grasses and wildflowers, and pinus aristata 'bristlecomb pine.' The extensive meadow is planted with wildflowers and grasses, including various aquilegia sps columbines; baptisia geranium; lupinus sps lupine; pulminaria sps lungwort; phlox and helianthus; festuca; chasmanthus northern seed oats; agropyron bluebunch wheatgrass.
PLANTING METHOD: Planted by hand.
SPECIAL FEATURES: The entire landscape is constructed over a massive slab of concrete and steel irrigation lines, polystytrene blocks, plastic drainage matting, elastomeric waterproofing, and inert lightweight substrate. The meadow is sustained by a thoroughly integrated irrigation system. Winner of the 2003 Green Roof Awards of Excellence.
GREEN BUILDING FEATURES: n/a
COST: n/a
WEIGHT: 278-513 lbs/s.f.
SUBMITTED BY: Zimmer Gunsul Frasca Partnership

21 | Osaka Municipal Central Gymnasium

BUILDING TYPE
Institutional
NEW/RETROFIT
New
GREEN ROOF TYPE
Intensive
GREEN ROOF SIZE
1,101,089 s.f.
PERCENTAGE OF ROOF COVERED
96%

Green Roof System Details

SYSTEM MANUFACTURER: n/a
MEMBRANE: Seam-welded stainless sheet
DRAINAGE: Granite
SOIL MEDIUM: Decomposed granite
SOIL DEPTH: 39.37 in. average
PLANTS: Lawn, shrubs, and trees
PLANTING METHOD: Planted by hand

SPECIAL FEATURES: A sprinkler system, a supply pipe, and a permeable drainage pipe are arranged in a pattern mesh over the entire roof.
GREEN BUILDING FEATURES: Natural lighting and ventilation system.
COST: $877.23/s.f.
WEIGHT: 1,507.2 lbs/s.f.
SUBMITTED BY: Nikken Sekkei

22 | Daimler Chrysler Complex

BUILDING TYPE
Mixed-use
NEW/RETROFIT
New
GREEN ROOF TYPE
Extensive and Intensive
GREEN ROOF SIZE
172,224 s.f.
PERCENTAGE OF ROOF COVERED
90%

Green Roof System Details

SYSTEM MANUFACTURER: Zinco, Enkadrain, and Eggers Soil
MEMBRANE: n/a
DRAINAGE: ZinCo FD20 40 60 systems. There is also a wire mat to allow air circulation between the insulation and the drainage. Drainage layer is filled with puffed shale.
SOIL MEDIUM: Expanded slate and pumice mix.
SOIL DEPTH: 4-23.6 in. Soil was blown up to the roof during the winter when the substrate was wet, which was messy and caused problems with the equipment and contactors.
PLANTS: Sedums, shrubs, 30-year-old trees
PLANTING METHOD: Hand-spread sedum cuttings; hand-planted shrubs and trees
SPECIAL FEATURES: Irrigation
GREEN BUILDING FEATURES: Rainwater toilets and rain barrels for irrigation. Residents have complained that there is too little recycled water for irrigation.
COST: n/a
WEIGHT: 20-61.4 lbs/s.f.
SUBMITTED BY: Marco Schmidt and Daniel Roehr

23 | Roppongi Hills

BUILDING TYPE
Mixed-use
NEW/RETROFIT
New
GREEN ROOF TYPE
Extensive and Intensive
GREEN ROOF SIZE
143,000 s.f.
PERCENTAGE OF ROOF COVERED
26%

Green Roof System Details

SYSTEM MANUFACTURER: Green Mass Damper Technology
MEMBRANE: Rubberized Asphalt
DRAINAGE: n/a
SOIL MEDIUM: *Various*: Normal soil, artificial lightweight aggregates, and sedum mats were all used in localized areas.
SOIL DEPTH: 1.17-46.8 in. The areas planted with sedum mats are the shallowest, at 1.17 in. The lightweight aggregate depth is 11.7 in.-31.2 in. Areas in which trees were planted in local soil are 46.8 in. deep.
PLANTS: Sedums, rice, edibles, and many other perennials and annuals.
PLANTING METHOD: Sedum mats and hand-planted plugs, bulbs, and root balls. Underground props stabilize the trees.
SPECIAL FEATURES: A drip irrigation system collects water in rain barrels and stores it in underground tanks in local areas. Some of the water is filtered through a pond and into the rice paddy field.
GREEN BUILDING FEATURES: Earthquake-resistant technology and a water recycling system.
COST: n/a
WEIGHT: 20 lbs/s.f.
SUBMITTED BY: Keiko Horioka, Mori Building Company

24 | Vastra Hamnen

BUILDING TYPE
Single-family residential
NEW/RETROFIT
New
GREEN ROOF TYPE
Extensive
GREEN ROOF SIZE
48,000 s.f. over 16 buildings
PERCENTAGE OF ROOF COVERED
60%

Green Roof System Details

SYSTEM MANUFACTURER: Veg Tech AB
MEMBRANE: Modified Bitumen
DRAINAGE: Roofs with a slope of less than 5° use Veg Tech's Drainfelt. Pitched roofs use the Hydrofelt system.
SOIL MEDIUM: Limestone peat, lava, and soil.
SOIL DEPTH: 1.4 in.
PLANTS: Sedum
PLANTING METHOD: Pre-fabricated Veg Tech/Xeroflor mats
SPECIAL FEATURES: n/a
GREEN BUILDING FEATURES: Active solar and open channel stormwater management.
COST: $6.5/s.f.
WEIGHT: 10 lbs/s.f.
SUBMITTED BY: Veg Tech AB

25 | GENO Haus

BUILDING TYPE
Office
NEW/RETROFIT
New and retrofit
GREEN ROOF TYPE
Extensive and intensive
GREEN ROOF SIZE
30,000 s.f.
PERCENTAGE OF ROOF COVERED
86%

Green Roof System Details

SYSTEM MANUFACTURER: Optima
MEMBRANE: Modified Bitumen
DRAINAGE: Optima Type Perl with a water-ponding system
SOIL MEDIUM: Optima Type I
SOIL DEPTH: 4-10 in.
PLANTS: *Intensive areas:* Low growing plants: Cerastium tomentosum; Hieracium pilosella; Calamagrostis acutiflora 'Karl Foerster'; Pinus mugo var. pumilio; Spirea japonica 'Little Princess'; Potentilla fruticosa; sedums. High growing plants: Amelanchier lamarcki; Prunus serrulata 'Kanzan'; Prunus subhirtella Autumnalis; Hibiscus syriacus; Buddleja alternifolia; Spirea arguta. *Windy areas:* Acer campestre; Cornus mas 'Otto Luyken.' *Shady areas:* Buxus sempervirens; Prunus laurocerasus 'Otto Luyken.'
SPECIAL FEATURES: Plants were selected for specific areas based on the environmental conditions of wind and sun exposure, and visibility. The intensive gardens were hand-planted on lower buildings, which are shaded by the taller buildings. Low-growing plants were selected to be visible from the conference room without blocking the view of the city. In total there are three intensive, one semi-intensive, and nine extensive gardens on the four-building complex.
GREEN BUILDING FEATURES: n/a
COST: Intensive: $14/s.f.; extensive: $4/s.f.
WEIGHT: 20 lbs/s.f.
SUBMITTED BY: Peter Philippi, Green Roof Service

26 | Schachermayer Company

BUILDING TYPE
Industrial
NEW/RETROFIT
Retrofit
GREEN ROOF TYPE
Extensive
GREEN ROOF SIZE
376,740 s.f.
PERCENTAGE OF ROOF COVERED
80%

Green Roof System Details

SYSTEM MANUFACTURER: Optigrün
MEMBRANE: Natural foil rubber

DRAINAGE: Optigrün System 10

SOIL MEDIUM: Optigrün Type M

SOIL DEPTH: 4 in.

PLANTS: Sedum

PLANTING METHOD: Hydro seed

SPECIAL FEATURES: Terrace with snack bar

GREEN BUILDING FEATURES: n/a

COST: n/a

WEIGHT: 24 lbs/s.f.

SUBMITTED BY: Edmund Maurer, City of Linz

27 | Augustenborg Botanical Garden

BUILDING TYPE
Institutional

NEW/RETROFIT
Retrofit

GREEN ROOF TYPE
Extensive and intensive

GREEN ROOF SIZE
95,000 s.f.

PERCENTAGE OF ROOF COVERED
93%

Green Roof System Details

SYSTEM MANUFACTURER: Veg Tech AB

MEMBRANE: Modified Bitumen

DRAINAGE: Various

SOIL MEDIUM: *Various*: Veg Tech light soil mix; gravel; crushed brick; rockwool; aquatop (a recycled foam material).

SOIL DEPTH: 1 in.-5.25 ft.

PLANTS: Various

PLANTING METHOD: Various

SPECIAL FEATURES: The roof is divided into 14 subsections. Below is an outline of their compositions. Roof B1 has three different slopes built for meteorology, water retention, run-off quality, and plant development testing. The entire roof is divided into 3.28-ft. (1 m) sections with different kinds of drainage materials replicated at irregular intervals. Materials include gravel, crushed brick, rockwool, aquatop (recycled foam material), and nothing. The drainage layer ranges from .4 in. to 1.18 in. deep. Pre-fabricated moss-sedum mats cover the drainage layers. The substrate is about 1.18 in. (3 cm) deep.

Roof B2 is divided into 9.6-ft. (3 m) sections to demonstrate how drainage layer affects plants. Prefabricated moss-sedum mats are used with a rockwool substrate kept in place with gravel.

C1 currently contains five inspiration gardens. Plot 1 has high bamboo sticks with climbers (Virginia creeper and Polygonatum). Soil layer 1.18-15.75 in. (3-40 cm) in mounds around the bamboo sticks. Ground cover is made up of sedum mats and large-leaved sedum plants. Plot 2 is a landscape with polystyrene-built hills up to 5.25 ft. (1.6 m), covered with prefabricated mats of grass and dry grassland flower species. The soil layer is almost 6 in. (15 cm). Plot 3 replicates a wooden veranda floor, surrounded by strict lavender hedges and large wooden boxes for flowers and herbs. The boxes also contain three connected ponds. The soil layer is made of 6-in. (15 cm) rockwool. Plot 4 is a waterscape made of a stream running over a slate "riverbed," surrounded by water vegetation and meadows. Soil layer ranges 0-6 in. deep. There is no drainage layer. Instead, water-retaining rubber-lined wood structures contain the water vegetation, and polystyrene forms the shape of the valley. Plot 5 is gravel, with no drainage layer, other than a small area of rockwool where small birch trees are planned. Gravel depths are 3.9-6 in. (10-15 cm).

C2 is divided into 10.8 x 10.8 s.f. (1 x 1 s.m.) plots of different rooftop-suitable plant species. Soil depth is 6 in. (15 cm), with no drainage layer.

D1 experiments with different methods of establishment: prefabricated mats, hand planted plugs, seeds, cuttings, and one plot left unplanted for spontaneous establishment. The drainage layer is made of Aquatop.

D2 is similar to C2, but the soil layer is a thinner .8-2 in. (3-5 cm). Sedums, other succulent species, and 19 species of moss are exhibited.

E1 is made of pre-fabricated sedum mats cut into star shapes and divided with gravel.

E2 is made of pre-fabricated mats on a small hill with a pond.

F1 is made of a very thin substrate layers and prefabricated sedum. Seagulls nest here. This will likely be the site for solar panel installation.

F3: Different colors of sedum are planted in gravel and crushed brick in wave shapes.

F5: This research site is experimenting with different soils and mixes. It is planted with extensive sedum mats.

G1 is divided into 3.28-ft. (1 m) sections, and experiments with an irregular pattern of soil mixtures, drainage layers, and establishment methods.

G2 exhibits extra-thin, lightweight layers of materials made by various green roof companies

G3 is a standard prefabricated mat plot.

GREEN BUILDING FEATURES: n/a

COST: $4/s.f. for extensive gardens

WEIGHT: 7-52 lbs/s.f.

SUBMITTED BY: Louise Lundberg, International Green Roof Institute (IGRI)

28 | Milwaukee Metropolitan Sewerage District

BUILDING TYPE
Office

NEW/RETROFIT
Retrofit

GREEN ROOF TYPE
Extensive

GREEN ROOF SIZE
3,800 s.f.

PERCENTAGE OF ROOF COVERED
40%

Green Roof System Details

SYSTEM MANUFACTURER: Weston Solutions Inc. and ABC Roofing Supply Inc.

MEMBRANE: Modified Bitumen

DRAINAGE: 2 x 6 perforated recycled plastic containers

SOIL MEDIUM: n/a

SOIL DEPTH: 4 in.

PLANTS: Native plants from the dry goat prairies of southwestern Wisconsin, including prairie drop seed, little bluestem, alum root, and wild columbine.

PLANTING METHOD: 2-year plugs were planted in the trays two weeks prior to installation. Native prairie seed collected from a bluff in southern Wisconsin was purchased to supplement the plugs. Approximately 10% of this seed was added to the trays when the roof initially installed in July 2003, and the balance planted in April 2004.

SPECIAL FEATURES: Flow-monitoring devices measure runoff.

GREEN BUILDING FEATURES: n/a

COST: n/a

WEIGHT: n/a

SUBMITTED BY: Stephen McCarthy, Milwaukee Metropolitan Sewerage District

29 | Montgomery Park Business Center

BUILDING TYPE
Office
NEW/RETROFIT
Retrofit
GREEN ROOF TYPE
Extensive
GREEN ROOF SIZE
30,000 s.f.: 20,000 s.f. on main building and 10,000 s.f. on warehouse
PERCENTAGE OF ROOF COVERED
18%

Green Roof System Details
SYSTEM MANUFACTURER: n/a
MEMBRANE: PVC
DRAINAGE: n/a
SOIL MEDIUM: Stalite and organic soil
SOIL DEPTH: 2.5-3 in.
PLANTS: Sedum acre Aureum 'Golden Stone Crop,' Sedum album Chloroticum, Sedum 'Blue Carpet,' Sedum cyaneum 'Rose Carpet,' Sedum dasyphyllum 'Saul Brothers,' Sedum dasyphyllum 'Blue Carpet,' Sedum ewersii 'Pink Stone Crop,' Sedum 'Jelly Bean,' Sedum kamtschaticum, Sedum kamtschaticum Variegatum, Sedum Pinifolium 'Blue Spruce,' Sedum sexangulare, Sedum spurium Fuldaglut, Sedum spurium 'White Form,' Sedum Weihenstephaner Gold, and Sempervivum sp. 'Hens & Chicks.'
PLANTING METHOD: Sedum, cuttings and seeds
SPECIAL FEATURES: Winner of the 2003 Green Roof Awards of Excellence
GREEN BUILDING FEATURES: Other green building features include: bioretention areas; asphalt paving recycling; 10,000 gallon rainwater storage cistern for toilet flushing, waterless urinals, and ice storage; operable windows with low e-glazing; energy-efficient lighting and elevator; raised floors for wiring; recycled carpet; workstations made of recycled paper, recycled wheat board, and sustainable ash trim. A smart building control system optimizes the use of the outside air for cooling, controls lighting based on daylight and utilizes high-efficiency boilers, chillers, and ice storage.
COST: n/a
WEIGHT: n/a
SUBMITTED BY: Katrin Scholz-Barth Consulting

30 | Somoval Garbage Treatment Plant

BUILDING TYPE
Industrial
NEW/RETROFIT
New
GREEN ROOF TYPE
Extensive
GREEN ROOF SIZE
165,000 s.f.
PERCENTAGE OF ROOF COVERED
n/a

Green Roof System Details
SYSTEM MANUFACTURER: Soprema
MEMBRANE: Elastomeric Bitumen Sopralène Flam Jardin
DRAINAGE: Flat area has no other drainage than substrate; sloped area uses a canalled geotextile.
SOIL MEDIUM: Flat area uses Sopraflor M 010 substrate; sloped area uses Sopraflor X 090
SOIL DEPTH: 2.4-2.8 in.
PLANTS: Sedum
PLANTING METHOD: Cuttings
SPECIAL FEATURES: n/a
GREEN BUILDING FEATURES: Energy efficiency and innovative garbage treatment.
COST: $2.20/s.f.
WEIGHT: 36.45lbs/s.f.
SUBMITTED BY: Francois LaSalle, Soprema

31| Valdemingómez Recycling Plant

BUILDING TYPE
Industrial
NEW/RETROFIT
New
GREEN ROOF TYPE
Extensive
GREEN ROOF SIZE
161,458.7 s.f.

PERCENTAGE OF ROOF COVERED
65%

Green Roof System Details
SYSTEM MANUFACTURER: AIMAD and ZinCo GmbH
MEMBRANE: PVC
DRAINAGE: Floradrain FD25
SOIL MEDIUM: Volcanic rock and compost
SOIL DEPTH: 2.8 in.
PLANTS: Sedum
PLANTING METHOD: Pre-fabricated sedum carpet
SPECIAL FEATURES: Drip irrigation
GREEN BUILDING FEATURES: Energy efficiency
COST: $30/s.m.
WEIGHT: n/a
SUBMITTED BY: Christof Brinkmann, Abalos & Herreros

32 | Ford Rouge Center, Truck Plant

BUILDING TYPE
Industrial
NEW/RETROFIT
Retrofit
GREEN ROOF TYPE
Extensive
GREEN ROOF SIZE
454,000 s.f.
PERCENTAGE OF ROOF COVERED
41%

Green Roof System Details
SYSTEM MANUFACTURER: Xero Flor America
MEMBRANE: Modified Bitumen
DRAINAGE: Xerodrain nylon mesh drainage layer and felt fleece water retention layer
SOIL MEDIUM: Permatil
SOIL DEPTH: 2 in.
PLANTS: Sedum 'Fulda Glow,' Sedum 'Diffusum,' Sedum acre, Sedum kamtschaticum, Sedum album reflexum, Sedum ellacombeanum, and Sedum 'Coccineum.'
PLANTING METHOD: Cuttings and seeds
SPECIAL FEATURES: Winner of the 2004 Green Roof Awards of Excellence
GREEN BUILDING FEATURES: n/a
COST: n/a
WEIGHT: 9.7lbs/s.f.
SUBMITTED BY: William McDonough + Partners

33 | Possmann Cider Company

BUILDING TYPE
Industrial
NEW/RETROFIT
Retrofit
GREEN ROOF TYPE
Intensive
GREEN ROOF SIZE
32,292 s.f.
PERCENTAGE OF ROOF COVERED
100%

Green Roof System Details

SYSTEM MANUFACTURER: n/a
MEMBRANE: One waterproofing membrane covered with a copper root repellant membrane
DRAINAGE: Standard porous mat
SOIL MEDIUM: None
SOIL DEPTH: n/a
PLANTS: Riparian vegetation patented using the logatainer and phytolyse method. The plant composition has changed since the original planting in 1991. Some have disappeared and other new species have established themselves. *Original plantings:* Carex acutiformis; Carex gracilis; Carex vulpine; Carex paniculata; Carex pseudocyperus; Carex nigra; Carex panicea; Juncus effuses; Juncus inflexus; Iris pseudacorus; Caltha palustris; Myosotis palustris; Scirpus lacustris; Lythrum salicaria; Scirpus silvaticus; Alisma plantago aquatica; Ranunculus repens; Potentilla reptans; Potentilla anserine; Glechoma hederacea. *Original plants remaining in 1999:* Carex acutiformis; Carex gracilis; Carex paniculata; Carex pseudocyperus; Juncus effusus; Caltha palustris; Potentilla anserina; Potentilla reptans; Glechoma hederacea. *New plants found in 1999:* Achillea ptarmica; Agrostis canina; Agrostis stolonifera; Betula pendula; Bidens tripartita; Calamagrostis epigejos; Epilobium parviflorum; Equisetum arvense; Festuca rubra; Filipendula ulmaria; Galium palustre; Juglans regia; Lactuca serriola; Lotus pedunculatus; Lycopus europaeus; Lysimachia vulgaris; Lythrum salicaria; Mentha aquatica; Mimulus guttatus; Myosotis scorpioides; Persicaria maculosa; Phragmites australis; Poa palustris; Ranunculus repens; Salix alba; Salix caprea; Schoenoplectus lacustris; Scirpus sylvaticus; Scutellaria galericulata; Solidago canadensis; Stachys palustris; Tanacetum vulgare;

Taraxcum officinale; Trifolium repens; Veronica beccabunga; Vicia tetrasperma.
PLANTING METHOD: Planted by hand
SPECIAL FEATURES: The roof is a vital part of the cider-making process.
GREEN BUILDING FEATURES: The green roof cools the water for fermentation and refrigeration.
COST: n/a
WEIGHT: n/a
SUBMITTED BY: Andreas Dietz, Werner Volkmar Possmann, and Ulrich Zens.

34 | John Deere Works

BUILDING TYPE
Industrial
NEW/RETROFIT
Retrofit
GREEN ROOF TYPE
Intensive
GREEN ROOF SIZE
450 s.f.
PERCENTAGE OF ROOF COVERED
65%

Green Roof System Details

SYSTEM MANUFACTURER: n/a
MEMBRANE: Modified Bitumen
DRAINAGE: Overflow system with flood control
SOIL MEDIUM: n/a
SOIL DEPTH: 2 in.
PLANTS: Wetland plants: sedges carex, rushes jucus, and irises, all grown to function without a soil base.
PLANTING METHOD: n/a
SPECIAL FEATURES: n/a
GREEN BUILDING FEATURES: A solar system fuels the computers and pumps.
COST: $50/s.f.
WEIGHT: 70lbs/s.f.
SUBMITTED BY: Hartmut Bauer, John Deere

35 | Zurich Main Station

BUILDING TYPE
Transportation hub
NEW/RETROFIT
New
GREEN ROOF TYPE
Semi-intensive
GREEN ROOF SIZE
107,640 s.f.
PERCENTAGE OF ROOF COVERED
100%

Green Roof System Details

SYSTEM MANUFACTURER: Sarna
MEMBRANE: Sarnafil TG 66-16
DRAINAGE: PE-foam pad tissue, SARNA and gravel
SOIL MEDIUM: Gravel and sand incorporated with recycled topsoil
SOIL DEPTH: 4-12 in.
PLANTS: Species to attract wild bees, grasshoppers, and spiders, including various sedum acre, album, hispanicum, reflexum, sexangulare, spurium, and seeds of tanacetum vulgare 'common tansy,' Centaurea sp. cornflower, allium schoenoprasum chive leaves, Echium vulgare viper bugloss, and Campanula sp.
PLANTING METHOD: Hand planted and sprayed seed
SPECIAL FEATURES: The roof is connected to the ground with a planted tower and fence to give access to insects that cannot fly.
GREEN BUILDING FEATURES: n/a
COST: $20lbs/s.f.
WEIGHT: 50lbs/s.f. and 150-200 lbs/s.f.
SUBMITTED BY: Tino Reinecke, Ernst Basler + Partners Ltd.

36 | Laban Dance Centre

BUILDING TYPE
Institution
NEW/RETROFIT
New
GREEN ROOF TYPE
Intensive rubble roof
GREEN ROOF SIZE
4,305.6 s.f.
PERCENTAGE OF ROOF COVERED
33%

Green Roof System Details
SYSTEM MANUFACTURER: Bespoke and Trocal
MEMBRANE: Trocal PVC
DRAINAGE: Geotextile membrane
SOIL MEDIUM: Rubble
SOIL DEPTH: 5.1-6 in.
PLANTS: No intentionally deposited seeds or plants
PLANTING METHOD: n/a
SPECIAL FEATURES: Designed specifically for black redstarts and brownfield invertebrates.
GREEN BUILDING FEATURES: n/a
COST: $3.30/s.f.
WEIGHT: n/a
SUBMITTED BY: Dusty Gedge, Livingroofs.org

37 | Rossetti Bau

BUILDING TYPE
Institution
NEW/RETROFIT
New
GREEN ROOF TYPE
Extensive and intensive
GREEN ROOF SIZE
13,500 s.f.
PERCENTAGE OF ROOF COVERED
90%

Green Roof System Details
SYSTEM MANUFACTURER: Biber Dach AG Basel
MEMBRANE: Modified Bitumen
DRAINAGE: Protection and moisture retention
SOIL MEDIUM: Soil from onsite
SOIL DEPTH: 3-20 in.
PLANTS: Achillea millefolium; Campanula rotundifolia; Cerastium sp.; Chrysanthemum leucanthemum; Clinopodium vulgare; Conyca; Crepis capilaris; Dianthus carthusianorum; Echium vulgare; Erigeron annuus; Euphrsia roskoviena; Globularia sp.; Hieracium pilosella; Lactuca serriola; Leontodon hispidus; Medicago lupulina; Meliolotus albus/indicus; Papaver rhoeas; Petroraghia saxifraga; Plantago lanceolata; Potentilla argentea; Prunella vulgaris; Rosmarinus officinalis; Salvia pratensis; Scarbiosa columbaria; Sedum acre; Sedum reflexum; Sedum sexangulare; and Silene otites.
PLANTING METHOD: Seeds
SPECIAL FEATURES: n/a
GREEN BUILDING FEATURES: n/a
COST: $1.50/s.f.
WEIGHT: 30 lbs/s.f.
SUBMITTED BY: Stephan Brenneisen

38 | Bird Paradise

BUILDING TYPE
Institutional
NEW/RETROFIT
New
GREEN ROOF TYPE
Extensive
GREEN ROOF SIZE
20,000 s.f.
PERCENTAGE OF ROOF COVERED
60%

Green Roof System Details
SYSTEM MANUFACTURER: n/a
MEMBRANE: Asphalt
DRAINAGE: n/a
SOIL MEDIUM: 60% sand and gravel, topped by 40% topsoil with stones and humus
SOIL DEPTH: 3.5 in.
PLANTS: Local dry grassland plants
PLANTING METHOD: Seeds
SPECIAL FEATURES: n/a
GREEN BUILDING FEATURES: n/a
COST: n/a
WEIGHT: n/a
SUBMITTED BY: Stephan Brenneisen

Green Roof System Details
SYSTEM MANUFACTURER: n/a
MEMBRANE: Rubberized Asphalt
DRAINAGE: 2.5 in. pumice gravel
SOIL MEDIUM: 40% pumice; 40% volcanic rock; 5% coconut fibers; 15% accumulated organic matter
SOIL DEPTH: 7 in.
PLANTS: Senecio praecox; Echeveria gibbiflora Furcraea; sp. Sedum moranense; Sedum griseum; Sedum luteoviride; Sedum stahli; Sedum praealtum; Agave sp.; Agave Americana; Agave tequilana; Agave kerchovei; Astrophytum myriostigma; Astrophytum ornatum; Coryphantha erecta; Coryphantha sp.; Echinocereus delaetii; Echinocereus pentalophus; Echinocereus pectinatus; Echinocactus platyacanthus; Escontria chiotilla; Ferocactus flavovirens; Ferocactus haematacanthus; Ferocactus hystrix; Ferocactus latispinus; Ferocactus robustus; Mammillaria elongata; Mamillopsis senilis; Myrtillocactus schenkii; Myrtillocactus geometrizans; Neobuxbaumia tetetzo; Nopalxochia phyllanthoides; Opuntia ficus- indica; Opuntia imbricata; Opuntia robusta; Opuntia sp; Pachycereus hollianus; Pachycereus mitzy; Pereskiopsis diguetii; Pilosocereus chysacanthus; Familia cactaceae; Opuntia microdasys; Borzicactus aureospinus; Mammillaria prolifera; Echinopsis sp.; Lovibia silvestrii; Cereus peruvianus; Opuntia cylindrical; Heliocereus especiosus; Ferocactus pilosus; Mammillaria haagea; Mammillaria carnea; Mammillaria karnisensis; Astrophytum capricorne; Astrophytum geometricans; Euphorbia milii; Euphorbia pseudocactus; Euphorbia obovalifolia; Staphelia variegata; Staphelia gigantean; Duvalia sulcata; Aloe variegata; Aloe vera; Haworthia fasciata; Haworthia cymbiformis; Faucaria felina; Kalanchoe sp.
PLANTING METHOD: Root transplantation; cuttings; seeds
SPECIAL FEATURES: 500 s.f. are designed as an organoponic system to grow plants in compost. 200 s.f. are used to make a compost botanical garden for habitat preservation; 600 s.f. grows evergreen.
GREEN BUILDING FEATURES: n/a
COST: $12.56/s.f.
WEIGHT: 50 lbs/s.f.
SUBMITTED BY: Raul Gomez Porras, CICEANA

39 | Centro de Información y Comunicación Ambiental de Norte América (CICEANA)

BUILDING TYPE
Residential multi
NEW/RETROFIT
Retrofit
GREEN ROOF TYPE
Extensive and intensive
GREEN ROOF SIZE
5,812.6 s.f.
PERCENTAGE OF ROOF COVERED
73%

40 | Orchid Meadow

BUILDING TYPE
Industrial
NEW/RETROFIT
Retrofit
GREEN ROOF TYPE
Intensive
GREEN ROOF SIZE
100,000 s.f.
PERCENTAGE OF ROOF COVERED
100%

Green Roof System Details
SYSTEM MANUFACTURER: n/a
MEMBRANE: Asphalt
DRAINAGE: 2 in. sand and gravel
SOIL MEDIUM: Topsoil and humus
SOIL DEPTH: 8 in.
PLANTS: Grassland
PLANTING METHOD: Naturally deposited seeds
SPECIAL FEATURES: Organic growth
GREEN BUILDING FEATURES: n/a
COST: n/a
WEIGHT: 100 lbs/s.f.
SUBMITTED BY: Stephan Brenneisen

Endnotes

Introduction

[1] Elizabeth Schmidt and Kevin Young, eds., *Poems of New York* (New York: Knopf Publishing Group, 2002).

From Grey to Green: Environmental Benefits of Green Roofs

[1] Community Cartography/New York City Planning Agency/Smart Growth America; F. Kaid Benfield, Matthew D. Raimi, and Donald D.T. Chen, *Once There Were Greenfields: How Urban Sprawl Is Undermining America's Environment, Economy and Social Fabric* (New York: Natural Resources Defense Council, 1999).

[2] Hashem Akbari et. al., eds., *Cooling Our Communities: A Guidebook on Tree Planting and Light-Colored Surfacing* (Berkeley, Calif.: Lawrence Berkeley National Laboratory, 1992), p. xvii.

[3] Cynthia Rosenzwieg and William D. Solecki, eds., *Climate Change and a Global City: The Potential Consequences of Climate Variability and Change—Metro East Coast* (New York: Columbia Earth Institute, 2001), p. 9.

[4] Akbari et al., *Cooling Our Communities*, p. 16.

[5] James Brooks, "Heat Island Tokyo is in Global Warming Vanguard." *New York Times*, August 1, 2002.

[6] Though superficially related to color, albedo takes into account the full spectrum of solar radiation, from infrared to ultraviolet (UV).

[7] Akbari et al., *Cooling Our Communities*, p. 45.

[8] Ibid.

[9] Ibid., p. 16.

[10] Ibid., p. 18; Rosenzweig and Solecki, *Climate Change and a Global City*, p. 144.

[10.5] *Cooling Our Communities*, p. 21.

[11] Ibid., p. 21.

[12] Rosenzweig and Solecki, *Climate Change and a Global City*, p. 136.

[13] Karen Liu and Bas Baskaran, "Thermal Performance of Green Roofs through Field Evaluation" (paper presented at Greening Rooftops for Sustainable Communities: The First North American Green Roof Infrastructure Conference, Awards and Trade Show, Chicago, IL, May 2003).

[14] Karen Liu, "A National Research Council Canada Study Evaluates Green Roof Systems' Thermal Performance," *Professional Roofing* (National Roofing Contractors Association, 2004).

[15] Brad Bass, Roland Stull, Scott Krayenjoff, and Alberto Martilli, "Modelling the Impact of Green Roof Infrastructure on the Urban Heat Island in Toronto," *Green Roof Infrastructure Monitor* 4, no. 1 (2002): pp. 2-3; Steven Peck, "Towards an Integrated Green Roof Evaluation for Toronto," *Green Roof Infrastructure Monitor* 5, no. 1 (2003), p. 4.

[16] American Rivers, Natural Resources Defense Council, and Smart Growth America, *Paving Our Way to Water Shortages: How Sprawl Aggravates the Effects of Drought* (August, 2002), www.smartgrowthamerica.org/DroughtSprawlReport09.pdf.

[17] New York City's wastewater treatment plants are designed to handle, on average, twice the dry weather flow for limited periods. "Existing Conditions: NYC Sewer System," *New York's Adult Mosquito Control Program* (March, 2003) http://www.ci.nyc.ny.us/html/doh/pdf/wnv/3g.pdf.

[18] U.S. Environmental Protection Agency, "Combined Sewer Overflow (CSO) Control Policy," *Federal Register* 59, no. 75 (April 19, 1994), p. 3.

[19] U.S. Environmental Protection Agency, "Combined Sewer Overflows: Demographics" (April, 2004), http://cfpub.epa.gov/npdes/cso/demo.cfm.

[20] Andrea Elliot, "Sewage Spill During the Blackout Exposed a Lingering City Problem," *New York Times*, August 28, 2003.

[21] Tom Liptan. City of Portland Department of Environmental Services, personal interview, March 17, 2003.

[22] Clean Ocean and Shore Trust (COAST), "NY-NJ COAST Issue Summary: Combined Sewer Overflows" (April, 2004), www.nynjcoast.org/ARGO/issues/issues.html.

[23] Amy Moran, Bill Hunt, Dr. Greg Jennings, "A North Carolina Field Study to Evaluate Greenroof Runoff Quantity, Runoff Quality, and Plant Growth" (paper number 032303 presented at meeting of the American Society of Agricultural Engineers, Las Vegas, Nevada, July 2003); J.C. Denardo, A.R. Jarrett, H.B. Manbeck, D.J. Beattie, and R.D. Berghage, "Stormwater Detention and Retention Abilities in Green Roofs"; Abrams, Hutchinson, Retzlaff, and Liptan, "Stormwater Monitoring Two Ecoroofs in Portland, Oregon," May 2003.

[24] R. P. Hosker and Steven E. Lindberg, "Atmospheric Deposition and Plant Assimilation of Gases and Particles," *Atmospheric Environment* 16, pp. 889-910.

[25] Brad Bass, et al., "Modeling the Impact of Green Roof Infrastructure on the Urban Heat Island in Toronto." *Green Roof Infrastructure Monitor* 4, no. 1 (2002), pp. 2-3; Steven Peck, "Towards an Integrated Green Roof Evaluation for Toronto," p. 4.

[26] Tom Liptan, "Planning, Zoning and Financial Incentives for Ecoroofs in Portland, Oregon" (paper presented at Greening Rooftops for Sustainable Communities, Chicago, IL, May 2003).

[27] Kevin Burke, "Green Roofs and Regenerative Design Practices: The Gap's 901 Cherry Project" (paper presented at Greening Rooftops for Sustainable Communities, Chicago, IL, May 2003).

[28] Eric Shriner, "Conservation Architecture: Endangered Plants on an Old Slaughterhouse Roof" (paper presented at Greening Rooftops for Sustainable Communities, Chicago, IL, May 2003).

[29] Stephen Brenneisen, "The Benefits of Biodiversity from Green Roofs: Key Design Consequences" (paper presented at Greening Rooftops for Sustainable Communities, Chicago, IL, May 2003).

[30] Kevin Burke, "Green Roofs and Regenerative Design Strategies" (paper presented at Greening Rooftops for Sustainable Communities, Chicago, IL, May 2003).

Municipal Case Study

[1] Jane Jacobs, "The Greening of the City," *New York Times Magazine*, May 16, 2004/ Section 6.

Imagining the City: Urban Ecological Infrastructure

[1] Kenneth Frampton, *Studies in Tectonic Culture: The Poetics of Construction in Nineteenth and Twentieth Century Architecture* (Boston: MIT Press, 2001).

[2] Arjun Appadurai, *Modernity at Large* (Minneapolis: University of Minnesota Press, 1996), p. 31.

[3] S. E. van der Leeuw, S.E. and C. Aschan, "A Long-Term Perspective on Resilience" (paper presented at workshop on "System Shocks - System Resilience," Sweden, May 2000).

[4] David Harvey, *Justice, Nature and the Geography of Difference* (Massachusetts: Blackwell Publishers, 1996), p. 429.

[5] This source reference concerns the concept and function of ecological patches in general rather than the dynamics of green roofs in particular. S. T. Pickett and P. S. White, *The Ecology of Natural Disturbance and Patch Dynamics.* (Orlando, FL: Academic Press, 1985).

Berlin: Green Roof Technology and Policy Development

[1] Jahrbuch Dachbegruenung, 2002

[2] A. Auhagen and R. Bornkamm, "Versuche zur BegrŸnung von Daechern mit Vegetationsplatten," *Das Gartenamt* 26, no. 3 (1977), pp. 144-147.

[3] See Appendix I.

[4] F. Darius and J. Drepper, "Oekologische Untersuchungen auf bewachsenen Kiesdaechern in West Berlin," Master thesis, Technical University of Berlin, 1983, p. 1- 72; and F. Darius. and J. Drepper, "Rasendaecher in West Berlin," *Das Gartenamt* 33 (1984), pp. 5; 309-315.

[5] H. J. Liesecke, "Mineralische Schuettstoffe zur Herstellung von Vegetationssubstraten fuer Dachbegruenungen" *Dach + Gruen* 3 (2002), pp. 8-18.

[6] M. Koehler, ed., *Fassaden- und Dachbegruenung*, Stuttgart: Ulmer, 1993), p. 329; M. Koehler and M. Schmidt, "Hof- , Fassaden- und Dachbegruenung," *Landschaftsentwicklung und Umweltforschung* 105 (1997), pp. 1-177.

[7] Koehler and Schmidt, "Hof- , Fassaden- und Dachbegruenung."

[8] Marco Schmidt, "Rainwater Harvesting in Germany: New concepts for reducing the consumption of drinking water, flood control, and improving the quality of surface water and the urban climate."

[9] In 1999, the Berlin Senate pledged to develop a citywide program for sustainable development as outlined in the 1992 Rio Summit's Local Agenda 21 program. All 23 boroughs of Berlin are expected to develop their own plans. The Prenzlauerberg program is of note because of its continuation of the Courtyard Greening grant.

[10] Fiebig and Krause, 1983

[11] ZVG, GALK, and FLL. "Dachbegrünung direkt und indirekt gefördert," *Stadt und Grün*, April 1997, pp. 224-225.

Bibliography for Berlin
BGL- Bundesverband Garten-, Landschafts- und Sportplatzbau e.V. 2002, Jahrbuch Dachbegrüning, Thalaker Median, Braunschweig, Germany, pp. 170-229.

Dreiseitl, Herbert. "Gestaltungselemente für (Regen-)Wasser; Das Urbane Gewässer am Potsdamer Platz," *fbr-Wasserspiegel* Vol. 4 (1998), pp. 10-11.

Fachvereinigung Bauwerksbegrünung e.V., *Förderung von Dachbegrünung durch eine "Gespaltende Abwassersatzung"* brochure, February, 2000.

Guha, Ramachandea. *Environmentalism; a global history*. (New Delhi: Oxford University Press. 2000).

Hämmerle, Fritz. "Der Markt für grüne Dächer wächst immer weiter." *Jahrbuch Dachbegrünung*. 2002.

Hämmerle, Fritz. "Dachbegrünungen rechnen sich." *Jahrbuch Dachbegrünung*. 2002.

Hendriksen, Ingrid. Wohnumfeldverbesserung auf Privaten Freiflächen in der Inennstadt am Beispiel des Hofbegrünungsprogrammes Berlin. Gekürzte Fassung. Arbeitsgrupe Umweltfreundliche Bauwesen for Senatsverwaltung für Bau- und Wohnungswesen. 1990.

Klemek, Christopher. "Urbanism as Reform: Modernist Planning and the Crisis of Urban Liberalism in Europe and North America, 1945-1975." Dissertation, University of Pennsylvania, 2004.

Kuckartz, Udo and Heiko Grunenberg. Umweltbewusstsein in Deutschland 2002; Ergebnisse einer repräsentativen Bevölkerungsumfrage. Bundesministerium für Umwelt, Naturschutz und Reaktorsicherheit. 2002.

Schmidt, MarCo and Katharina Teschner. "Kombination von Regenwasserbewirtschaftungsmaßnahmen: Ergebnisse der Voruntersuchungen für das Projekt Potsdamer Platz- Teil 1: Stoffrückhalt extensiever," *Dachbegrünung gwf-Wasser/Abwasser* 141, no. 10 (2002).

ZVG, GALK, and FLL. "Dachbegrünung direkt und indirekt gefördert." *Stadt und Grün* April 1997.

Tokyo: Cooling Rooftop Gardens

[1] Figure 0.6 C comes from the Intergovernmental Panel on Climate Change Third Assessment Report, http://www.ipcc.ch/.

[2] Mean annual temperatures in downtown Tokyo have increased 5.2°F (3°C), while temperatures in New York City have increased only 2.88°F (1.6°C).

[3] Unknown, "Japanese Ministries to deal with 'Heat Island' Phenomenon," *Space Daily*. September 3, 2002, http://www.spacedaily.com/2002/020903114238.la4m5hpo.html.

[4] James Brooke, "'Heat Island' Tokyo is in Global Warming's Vanguard," *New York Times*. August 13, 2002.

[5] Ibid.

[6] Ibid.

[7] Unknown, "Japanese Ministries to deal with 'Heat Island' Phenomenon."

[8] Staff Writer, "Interest Growing in Roof Gardens," *The Yomiuri Shimbun*, March 19, 2002.

[9] Unknown, "Tokyo turns Rooftops into Gardens," *MSNBC News*, August 22, 2002, http://www.voy.com/61461/145.html.

[10] Tokyo Metropolitan Government, "Rooftop Greenery Measures" *The Environment in Tokyo 2002*, http://www.kankyo.metro.tokyo.jp/kouhou/englsih2002/index.html.

[11] Ibid.

London: The Wild Roof Renaissance

[1] *London Biodiversity Partnership The Action*, Volume 2 of the London Biodiversity Action Plan (London: London Biodiversity Partnership, 2001, 2002).

[2] S. Brownlie, "Roof gardens; a review," *Urban Wildlife Now* 7, Nature Conservancy Council, Peterborough, 1990; J. Johnston, and J. Newton, *Building Green: A guide to using plants on roofs, walls and pavements* (London: London Ecology Unit, 1993); J. Johnston, "Roof-top Wildlife," *Enact* 3 (1995), pp. 19-22.

[3] Urban Task Force, *Towards an Urban Renaissance*, (London: E&FN Spon, 1999).

[4] Office for National Statistics, General Register Office for Scotland, and Northern Ireland Statistics and Research Agency, "London Largest EU City" in British Census 2001. Published on 25 June 2003, http://www.statistics.gov.uk/cci/nugget.asp?id=384

[5] Department of the Environment, Transport and the Regions, *Our Towns and Cities: The future - delivering an urban renaissance*, (London: DETA, 2000).

[6] Matthew Frith, "Pretty vacant?," *Spaces & Places* 4 (2003), Urban Parks Forum, Reading, pp. 5-7.

[7] J. Scholfield, and M. Waugh, eds., *Brownfield?, Greenfield? The threat to London's unofficial countryside*, (London: London Wildlife Trust, 2003).

[8] M. Frith, P. Sinnadurai, and D. Gedge, *Black redstart – an advice note for its conservation in London* (London: London Wildlife Trust, 1999).

[9] Department for the Environment, Food, and Rural Affairs, *Working with the Grain of Nature: A Biodiversity Strategy for England*, (London: DEFRA, 2002).

[10] Department for the Environment, Food, and Rural Affairs, *Achieving a Better Quality of Life: Review of Progress Towards Sustainable Development*, Government Sustainable Development Report 2002, (London: DEFRA, 2003).

[11] DEFRA, *Working with the Grain of Nature*.

[12] The best known of these are outside London: the Willis Faber & Dumas building in Ipswich (built 1971, and recently listed by English Heritage), Gateway House in Basingstoke (built 1976), and the Eden Project near St. Austell (built 2001).

[13] Forschungsgesellschaft Landschaftentwicklung Landschaftbau e. V., *Guidelines for the Planning, Execution and Upkeep of Green-Roof Sites (English version)* (Bonn: FLL, 1995).

[14] Corporation of London and British Council for Offices, *Green Roofs: research advice note* (London: British Council for Offices, 2003).

[15] Greater London Authority, *Green Light to Clean Power: The Mayor's Energy Strategy* (London: Greater London Authority, 2004).

[16] Greater London Authority, *The London Plan; The Spatial Development Strategy for Greater London* (London: Greater London Authority, 2004).

Portland: A New Kind of Stormwater Management

[1] Dawn Hottenroth, "Using Incentives and Other Actions to Reduce Watershed Impacts from Existing Development," paper for City of Portland Bureau of Environmental Services, www.cleanrivers-pxd.org.

[2] Tom Liptan, City of Portland Department of Environmental Services, personal interview, March 17, 2003.

[3] Tom Liptan, "Planning, Zoning and Financial Incentives for Ecoroofs in Portland, Oregon" (paper presented at Greening Rooftops for Sustainable Communities, Chicago, IL, May 2003), p.6.

[3.5] Hutchinson, Abrams, Retzlaff, Liptan, "Stormwater Monitoring in Portland Oregon," May 2003.

[4] Ibid.

[5] Dawn Hottenroth, "Using Incentives and Other Actions to Reduce Watershed Impacts from Existing Development."

[6] Tom Liptan, "Planning, Zoning and Financial Incentives for Ecoroofs in Portland, Oregon," p. 6.

[7] Anthony Ray, City of Portland Department of Sustainable Development and Ecoroofs Everywhere, personal interview, May 17, 2003.

Acknowledgements
There are numerous people in Portland that have assisted with the advancement of the ecoroof program. A few who didn't roll their eyes (too much) when they first heard the idea were: Dean Marriott (Director of BES), Linda Dobson (former assistant to Dean), City Commissioners Dan Saltzman and Erik Sten, the City Council, Patrice Mango, Diana Hinton, Dave Kliewer (Environmental Services staff), Ed McNamara, and John Mason.

Chicago: Towards a New Standard of Green Building
[1] Kimberly A. Gray and Mary E. Finster, "The Urban Heat Island, Photochemical Smog, and Chicago: Local Features of the Problem and Solution," Atmospheric Pollution Prevention Division, p. 90.

[2] City of Chicago. Monitoring data available at http://www.cityofchicago.org.

[3] The energy code essentially adopts the standards of the American Society of Heating, Refrigerating and Air Conditioning Engineers.

Toronto: A Model for North American Infrastructure Development
[1] City of Toronto. *The Official Plan for the City of Toronto.* http://www.city.toronto.on.ca/torontoplan/774.htm.

[2] Brad Bass, Chris Callaghan, Monica Kuhn, and Steven Peck, *Green Backs from Green Roofs: Forging a New Industry in Canada* (Ottawa: CMHC-SCHL Canada, 1998).

[3] Flynn Canada, IRC Building Sciences, Soprema, Garland Canada, Sheridan Nurseries, and Semple Gooder.

[4] *Green Roof Infrastructure Monitor* 2, no.1 (spring 2000), pp.1-2.

[5] Ibid.

[6] The study used a mesoscale atmospheric model, to calculate the urban heat island effect reduction based on the creation of 65 million square feet of green roofs. The green roof prototype was a generic 6 in (15 cm) at the estimated cost of $45.5 million (CDN). Brad Bass, Roland Stull, Scott Krayenjoff, and Alberto Martilli, "Modelling the Impact of Green Roof Infrastructure on the Urban Heat Island in Toronto." *Green Roof Infrastructure Monitor* 4, no. 1 (2002), pp. 2-3.

[7] Steven Peck, "Towards an Integrated Green Roof Evaluation for Toronto." *Green Roof Infrastructure Monitor* 5, no. 1 (2003), p. 4.

[8] Ibid.

[9] Jill Killeen, Marketing Department at Fairmont Waterfront in Vancouver, personal interview, May 27, 2004.

[10] See paper presented by S. Peck and I. Wieditz at Greening Rooftops for Sustainable Communities Conference, Chicago (May 2003) describing the process of market transformation in detail, www.greenroofs.org.

New York: Greening Gotham's Rooftops
[1] Cynthia Rosenzweig and William Solecki, eds., *Climate Change and a Global City: The Potential Consequences of Climate Variability and Change – Metro East Coast.* Report for the U.S. Global Climate Change Research Program, National Assessment of the Potential Consequences of Climate Variability and Change in the Unites States (New York: Columbia Earth Institute, 2001).

[2] New York City Department of Environmental Protection, "2001 NY Harbor Water Quality Report" and "Summaries of NYC Sewage Plant Info on Dry Flow and Capacity."

[3] Andrea Elliot, "Sewage Spill During Blackout Exposes a Lingering Problem," *New York Times*, August 28, 2003.

[4] David Owen, "Concrete Jungle," *The New Yorker*, 2003.

[5] "Community Cartography, Land and Area Building Analysis for Five Boroughs of New York," commissioned by Earth Pledge, 2002.

[6] Unpublished Gaia Institute Study on South Bronx Stormwater Sewer Shed, 2004.

[7] Jane Jacobs, "The Greening of the City," *New York Times Magazine*, May 16, 2004.

Author Biographies

Colin Cheney is the director of the Earth Pledge Green Roofs Initiative. Since 2002, Cheney has overseen the range of education, facilitation, policy, and research activities supporting Earth Pledge's mission of solving urban environmental and human health problems through green roof infrastructure development. He co-edited "Green Roofs in the New York Metro Region: Research Report One of the New York Ecological Infrastructure Study," a summary of the findings of the multidisciplinary research project investigating the impacts, benefits, and costs of citywide green roof development, published in 2004. Cheney previously coordinated the environmental education initiative of The Rhode Island River of Words Project, a watershed-based ecology and arts program. He holds a degree in Environmental Studies from Brown University.

Mathew Frith is the landscape regeneration manager for Peabody Trust, where he works to develop and implement an ecology strategy for the Trust's housing estates across London. He became involved in the field of urban ecology in 1987, first with the London Ecology Unit, and later with the London Wildlife Trust. Frith joined English Nature in 2000 as the urban adviser. His work focused on a number of initiatives, including green spaces, cemeteries, and green roofs. He has a degree in zoology from the University of Exeter and is currently the Chair of the London Biodiversity Partnership's Built Structures Working Group, Vice-Chair of London Wildlife Trust, a CABE Space adviser, and a member of the UK MAB Urban Forum and the Institute of Ecology and Environmental Management.

Dusty Gedge is one of the leading advocates for green roofs in the United Kingdom. Gedge became involved in green roof development as a result of his work with protected urban birds, particularly the black redstart. He directs the Black Redstart Action Plan of the London Biodiversity Partnership (www.lbp.org.uk), which has led to the proposed establishment of over 2 million square feet of green roof redstart habitat. Gedge is a leading consultant on green roof design and cost/benefit analysis. He has also raised over $100,000 (£50,000) for the first major study of green roofs and biodiversity to be conducted outside of Switzerland, to which he is a supervisor and advisor. He is currently working to form an independent green roof organization (www.livingroofs.org.uk) with other like-minded people in the UK.

Leslie Hoffman has been the executive director of Earth Pledge since 1994. After graduating from Colorado College in 1979 with a degree in architecture and design, she went on to design and build environmentally friendly buildings in Maine for ten years. Long a pioneer of sustainable architecture and agriculture, Leslie developed three major initiatives at Earth Pledge that guide the organization's work: Farm to Table, Green Roofs, and Waste=Fuel Initiatives, all of which deliver viable models to government, industry, and communities through demonstration, education, and research. The organization's innovative work has received accolades from *The New York Times*, *The Wall Street Journal*, National Public Radio, and numerous industry and trade publications. In 2004, Earth Pledge was named an Environmental Achiever by the US Environmental Protection Agency, and was selected as an Environmental Champion by *Interiors and Sources* magazine.

Melissa Keeley is a US National Science Foundation Graduate Research Fellow, and is conducting her doctoral research at the Technical University of Berlin, Germany. Her research focuses on the transferability of stormwater management technologies from Europe to North America. She is examining the technical adaptations necessary to implement techniques including green roofs, porous pavers, and infiltration systems in North America, as well as the applicability of European public policy instruments that support these technologies. Keeley has worked as a stream restoration ecologist in greater Philadelphia. She completed her M.S. at the University of Washington's Center for Urban Horticulture in Seattle, Washington.

Manfred Koehler has been a professor of landscape ecology at the University of Applied Sciences Neubrandenburg in Mecklenburg-Vorpommern, Germany, since 1994. Koehler has conducted research on green roofs, green façades, and vegetation in the urban environment for over twenty years. His research interests include the plant ecology and ecological function of green roofs, and much of his work has focused on the green roofs of Berlin. Koehler is the author of numerous publications, including one of the first green roof and green façade instructional volumes, published in 1993 by Ulmer Press in Stuttgart. He is a member of the German Landscape Architectural Organization (BDLA) and, since 1995, has served as a member of the green roof working group of the German Federation of Landscaping (FLL), contributing to the codification of FLL green roof standards.

Tom Liptan is a landscape architect and stormwater specialist with the City of Portland Bureau of Environmental Services Sustainable Stormwater Management Program. He has been internationally recognized for his work on ecoroofs, water gardens, and other sustainable site design techniques. For the past eleven years, he has worked on the development of new site and building design/planning/management approaches, ecoroof research and program development, implementation of demonstration projects with public and private partners, stormwater monitoring, city code modifications, and education and outreach. He is co-author of the chapter "Stormwater Gardens" in the *Handbook of Water Sensitive Planning and Design*, by Robert France, published by Lewis Publishers in 2002.

William McDonough is an architect, founding principal of William McDonough + Partners Architecture and Community Design, and cofounder, with Dr. Michael Braungart, of McDonough Braungart Design Chemistry, a product and systems development firm. From 1994 to 1999, he served as Dean of the School of Architecture at the University of Virginia. In 1996, he received the Presidential Award for Sustainable Development, the highest environmental honor given by the United States, and in 1999 *Time* magazine recognized him as a "Hero for the Planet," stating that "his utopianism is grounded in a unified philosophy that—in demonstrable and practical ways—is changing the design of the world." He writes frequently on architecture and design for journals and magazines and is the co-author, with Michael Braungart, of *Cradle to Cradle: Remaking the Way We Make Things*, published in 2002 by North Point Press.

Takehiko Mikami has been a professor of climatology at Tokyo Metropolitan University since 1991. He completed his PhD at Tokyo University in 1977. His research, in collaboration with the Tokyo government, focuses on Tokyo's urban heat island. He is a member of the Japanese Government Committee on the Urban Heat Island. He is also a full member of the Commission on Climatology, and the International Geographical Union.

Steven W. Peck is the founder and executive director of Green Roofs for Healthy Cities, an association of public and private organizations dedicated to developing a multi-million-dollar green roof industry in North America. Peck has written and lectured widely on overcoming barriers to sustainable community development and green roof infrastructure implementation. Green Roofs for Healthy Cities provides education, training, and advocacy services to support government evaluation of and investment in green roof technology.

Katrin Scholz-Barth is a nationally recognized expert in green roof technology. Her work demonstrates that green roofs are an integral and functional building element that protects watersheds while enhancing urban biodiversity and quality of life. Prior to starting her own business, Katrin Scholz-Barth Consulting, she was Director of Sustainable Design for the HOK Planning Group, a business unit of Hellmuth, Obata, and Kassabaum (HOK). She was the principle designer and consultant for the green roof at the Montgomery Park Business Center in Baltimore, which received the Green Roof Award of Excellence in 2003, and has worked on thirteen projects in the United States. She is also an adjunct professor at the University of Pennsylvania's School of Design. Scholz-Barth is trained in masonry and bricklaying and received her Masters of Science in civil and environmental engineering from the University of Rostock, Germany. Scholz-Barth has co-authored a book on green roofs to be published by Wiley and Sons in 2005.

Ed Snodgrass is the owner and president of Emory Knoll Farms and Green Roof Plants, which has grown and distributed plants for projects in at least twenty states, the District of Columbia, Canada, Hong Kong, and Japan. By 2004, he had supplied plants for over 300,000 square feet of green roof and expects the sum to double by 2005. Snodgrass is a fifth-generation farmer and nurseryman specializing in plant stocks for green roofs. He is a member of the Maryland Nurserymen's Association, The Royal Horticulture Society, The Scottish Rock Garden Society, The Sedum Society, and The International Plant Propagators Society.

Joel Towers is the first Director of Sustainable Design and Urban Ecology at Parsons School of Design, New School University, and a partner in the office of SR+T Architects. The Parsons program is a trans-disciplinary undergraduate and graduate course of study that explores sustainability, urbanity, and ecology through art and design. Linked to the emerging environmental studies initiative across New School University, it provides a unique opportunity for students and faculty to develop socionatural models for the 21st century. A practicing architect, Towers holds a Masters Degree in Architecture from Columbia University and has served on the faculties of both City College and Columbia. He was the project director for The Hannover Principles: Design for Sustainability in the office of William McDonough Architects prior to co-founding SR+T Architects in 1992 with Karla Rothstein.

Acknowledgements

Green Roofs: Ecological Design and Construction was a collaborative effort. The building case studies were written by Earth Pledge, with information from original submissions and interviews. Every reasonable effort has been made to verify information. The editorial team apologizes for any errors or omissions in these narratives.

The editorial team thanks all those who donated their time and effort to create *Green Roofs: Ecological Design and Construction*, particularly: the City of Chicago, Matthew Frith, Dusty Gedge, Melissa Keeley, Manfred Kohler, Tom Liptan, William McDonough, Takehiko Mikami, Steven Peck, Katrin Scholz-Barth, and Joel Towers. In addition, we would like to acknowledge Jean Gardner for her contributions to this publication.

Green Roofs: Ecological Design and Construction benefited from the assistance of many firms, institutions, and manufacturers. For their enormous devotion of time and materials, the editorial team would like to thank all those individuals and organizations who submitted materials, photographs, and information for the book, particularly: Stephan Brenneisen; Heidrun Eckert, ZinCo GmbH; Mark Farina, City of Chicago; the staff of Green Roofs for Healthy Cities; G. Kadas; Haven Kiers; Kevin Laberge, City of Chicago; Edmund Mauer, City of Linz; William McDonough + Partners; Peter Philippi, Green Roof Service LLC; the City of Portland; Daniel Roehr; Marco Schmidt; Ed Snodgrass, Green Roof Plants; Rivkah Walton, Roofscapes Inc.; Kimberly Worthington, City of Chicago; and Ulrich Zens.

Finally, we would like to thank Schiffer Publishing for the opportunity to showcase our vision of ecological green roofs. This experience has enhanced our knowledge of the state of green roof development throughout the world, and has raised the bar of our expectations for the potential of creating ecological green roofs.

Front Cover: ACROS Fukuoka. Architect: Emilio Ambasz & Associates. Architect of Record: Nihon Sekkei and Takenaka Corporation. Construction Manager: Takenaka Corporation. Photographer: Hiromi Watanabe.

Contents Page (clockwise): Heinz 57 Center, Courtesy of Roofscapes Inc.; Monthyon Garbage Treatment Center, Courtesy of Soprema; sedum, Emory Knoll Farms.

Building Case Studies (clockwise): Vastra Hamnen, Courtesy of Johan Thiberg; Milwaukee Metropolitan Sewerage District, Courtesy of Bob Kuehn; CICEANA, Courtesy of CICEANA; Schachermayer Company, Courtesy of Magistrate of Linz.

Municipal Case Studies (clockwise for two pages): Earth Pledge Kitchen Garden, Courtesy of Earth Pledge; Chicago City Hall, Courtesy of Conservation Design Forum; St. Luke's Science Center Healing Garden, Courtesy of St. Luke's International Hospital; Toronto City Hall, Courtesy of Green Roofs for Healthy Cities; One Waterfront Place, Courtesy of BOORA Architects; Berlin, aerial view, Courtesy of Manfred Kohler.

Page 157: Courtesy of Hiromi Watanabe/Emilio Ambasz & Associates

Page 158: Courtesy of Christian Kandzia

Back Cover (clockwise): Chicago City Hall, Courtesy of Conservation Design Forum; Schachermayer Company, Courtesy of the Magistrate of Linz; Ford Rouge Center, Truck Plant, Courtesy of Ford Photographic/ William McDonough + Partners